Oliver Kiowski

Spectroscopy of Individual Single-Walled Carbon Nanotubes

Oliver Kiowski

Spectroscopy of Individual Single-Walled Carbon Nanotubes

and their Synthesis via Chemical Vapor Deposition

Südwestdeutscher Verlag für Hochschulschriften

Impressum/Imprint (nur für Deutschland/only for Germany)
Bibliografische Information der Deutschen Nationalbibliothek: Die Deutsche Nationalbibliothek verzeichnet diese Publikation in der Deutschen Nationalbibliografie; detaillierte bibliografische Daten sind im Internet über http://dnb.d-nb.de abrufbar.
Alle in diesem Buch genannten Marken und Produktnamen unterliegen warenzeichen-, marken- oder patentrechtlichem Schutz bzw. sind Warenzeichen oder eingetragene Warenzeichen der jeweiligen Inhaber. Die Wiedergabe von Marken, Produktnamen, Gebrauchsnamen, Handelsnamen, Warenbezeichnungen u.s.w. in diesem Werk berechtigt auch ohne besondere Kennzeichnung nicht zu der Annahme, dass solche Namen im Sinne der Warenzeichen- und Markenschutzgesetzgebung als frei zu betrachten wären und daher von jedermann benutzt werden dürften.

Coverbild: www.ingimage.com

Verlag: Südwestdeutscher Verlag für Hochschulschriften GmbH & Co. KG
Dudweiler Landstr. 99, 66123 Saarbrücken, Deutschland
Telefon +49 681 37 20 271-1, Telefax +49 681 37 20 271-0
Email: info@svh-verlag.de

Approved by: Karlsruhe, KIT, Diss., 2008

Herstellung in Deutschland:
Schaltungsdienst Lange o.H.G., Berlin
Books on Demand GmbH, Norderstedt
Reha GmbH, Saarbrücken
Amazon Distribution GmbH, Leipzig
ISBN: 978-3-8381-2930-3

Imprint (only for USA, GB)
Bibliographic information published by the Deutsche Nationalbibliothek: The Deutsche Nationalbibliothek lists this publication in the Deutsche Nationalbibliografie; detailed bibliographic data are available in the Internet at http://dnb.d-nb.de.
Any brand names and product names mentioned in this book are subject to trademark, brand or patent protection and are trademarks or registered trademarks of their respective holders. The use of brand names, product names, common names, trade names, product descriptions etc. even without a particular marking in this works is in no way to be construed to mean that such names may be regarded as unrestricted in respect of trademark and brand protection legislation and could thus be used by anyone.

Cover image: www.ingimage.com

Publisher: Südwestdeutscher Verlag für Hochschulschriften GmbH & Co. KG
Dudweiler Landstr. 99, 66123 Saarbrücken, Germany
Phone +49 681 37 20 271-1, Fax +49 681 37 20 271-0
Email: info@svh-verlag.de

Printed in the U.S.A.
Printed in the U.K. by (see last page)
ISBN: 978-3-8381-2930-3

Copyright © 2011 by the author and Südwestdeutscher Verlag für Hochschulschriften GmbH & Co. KG and licensors
All rights reserved. Saarbrücken 2011

Abstract

In the course of this thesis, a thermal chemical vapor deposition (CVD) reactor was designed, built and used to grow vertically and horizontally aligned carbon nanotube arrays. The as-grown nanotubes were investigated on a single tube level using near-infrared photoluminescence (PL) microscopy as well as Raman, atomic force and scanning electron microscopy (SEM). For photoluminescence excitation (PLE) spectroscopy of individual, semiconducting single-walled carbon nanotubes (SWNTs), a specialized PL set-up was built allowing completely automated measurements within an excitation wavelength range of ~ 600-$1000\,\text{nm}$ (employing 3 tunable lasers) and PL imaging with a spatial resolution down to $400\,\text{nm}$.

The PL of as-grown, suspended nanotubes was compared to that of SWNTs in $D_2O/$ surfactant dispersions as well as to SWNTs embedded in organic solvents. The observed shifts of the characteristic optical transition energies E_{11} (emission) and E_{22} (excitation) were attributed to changes in the nanotube surroundings. These result in different dielectric screening of electronic excitations (excitons) in SWNTs.

We also compared the PL of CVD-grown SWNTs (air-suspended and surface-bound) to that of deposited individual surfactant-coated ones at temperatures down to $4\,\text{K}$ and observed PL intermittency and spectral diffusion at low temperatures only for the latter.

PLE spectra of air-suspended, CVD-grown SWNTs and of dispersions enriched in a few chiralities enabled the first direct observation of two weakly emissive excitonic states below the lowest optically active E_{11} exciton. Energy separations and intensities of these states relative to the main E_{11} PL emission peak are discussed.

Our PL set-up was sensitive enough to detect the weak PL of as-grown SWNTs on dielectric surfaces—in a configuration relevant for many proposed SWNT-based opto- and electronic devices. This was used to image ultralong CVD-grown nanotubes on Si/SiO_2, to determine their chirality and to check structural integrity along the nanotube length. Furthermore, ultralong SWNTs were manipulated (moved, bent and fractured) employing an atomic force microscope. We demonstrate that PLE microscopy is a powerful method to detect residual uniaxial and torsional strain in SWNTs. A peculiar PL behavior was found at fractured nanotube sites.

Finally, a new approach to determine relative abundances and PL quantum yields of semiconducting SWNTs in dispersions is presented, which is based on statistical counting of individual nanotubes by means of PL spectroscopy.

Zusammenfassung

Einzelmolekülspektroskopie an Kohlenstoffnanoröhren und deren Herstellung mittels chemischer Gasphasenabscheidung

Im Rahmen dieser Arbeit wurde eine Apparatur zur chemischen Gasphasenabscheidung entwickelt, aufgebaut und dazu verwendet, vertikal und horizontal ausgerichtete Anordnungen aus Kohlenstoffnanoröhren wachsen zu lassen. So hergestellte, einzelne Nanoröhren wurden sowohl mittels eines Photolumineszenzmikroskops (PL Mikroskop), als auch mit Raman-, Rasterkraft- und Rasterelektronenmikroskopie untersucht. Für die Photolumineszenz-Anregungs-Spektroskopie (PLE Spektroskopie) an einwandigen Kohlenstoffnanoröhren (SWNTs) wurde ein PL Lasermikroskop konstruiert, das automatisierte Messungen in einem Anregungswellenlängenbereich von ~ 600-1000 nm (mittels 3 durchstimmbarer Laser) und die Aufnahme von PL Bildern mit einer räumlichen Auflösung von bis zu 400 nm gestattete.

Die PL von direkt gewachsenen, frei hängenden SWNTs wurde mit der von SWNTs in D_2O/Tensid Dispersionen und ebenso mit der von SWNTs in organischen Lösungsmitteln verglichen. Die beobachteten Verschiebungen der charakteristischen optischen E_{11} Emissions- und E_{22} Anregungsenergien kommen durch Veränderungen in der Umgebung der Nanoröhren zustande. Dadurch ergibt sich eine unterschiedliche dielektrische Abschirmung der elektronischen Anregungen (Exzitonen).

Ebenso wurde die PL von direkt gewachsenen SWNTs (frei hängend und auf der Oberfläche liegend) mit der von einzelnen, aus D_2O/Tensid Dispersionen heraus auf einer Oberfläche abgeschiedenen SWNTs bei Temperaturen bis zu 4 K verglichen. Nur letztere zeigen bei tiefen Temperaturen PL-Blinken und spektrale Diffusion.

PLE Spektren von frei hängenden SWNTs und von Dispersionen, in denen einzelne Chiralitäten angereichert waren, ermöglichten die erste direkte Beobachtung von zwei tiefliegenden Exzitonenzuständen unterhalb des tiefsten optisch erlaubten E_{11} Exzitons. Die energetische Position dieser Zustände, sowie deren Intensität im Vergleich zu der E_{11} Emissionslinie werden diskutiert.

Das PL Mikroskop war empfindlich genug, um die schwache PL von direkt gewachsenen SWNTs auf dielektrischen Oberflächen zu detektieren. Extrem lange, mittels CVD auf Si/SiO_2 synthetisierte SWNTs konnten so abgebildet, ihre Chiralität bestimmt und die strukturelle Integrität entlang der Aufrollachse überprüft werden. Darüber hinaus wurden solche Nanoröhren mittels Rasterkraftmikroskopie manipuliert (verschoben, verbogen und zerrissen). Wir zeigen, dass PLE Mikroskopie eine leistungsstarke Methode ist, um noch vorhandene axiale und torsionale Dehnung in SWNTs nachzuweisen. An gebrochenen Stellen wurde bei Nanoröhren ein auffälliges PL Verhalten gefunden.

Schließlich wird ein neues Verfahren vorgestellt, mit dem relative Konzentrationen sowie Quantenausbeuten dispergierter, halbleitender SWNTs durch statistisches Zählen von PL Signalen bestimmt werden können.

Contents

1	**Introduction**	**1**
2	**Carbon Nanotube Basics**	**3**
2.1	Geometric Structure	3
2.2	Electronic Band Structure	6
2.3	Electronic Density of States	11
2.4	Optical Properties	16
	2.4.1 Overview	16
	2.4.2 Selection Rules	17
	2.4.3 Photoluminescence and Absorption Spectroscopy	21
	2.4.4 Excitons in Carbon Nanotubes	23
	2.4.5 Raman Spectroscopy	28
	2.4.6 Conclusion to Spectroscopic Methods	31
2.5	Synthesis Methods of Carbon Nanotubes	33
	2.5.1 Arc Discharge	33
	2.5.2 Pulsed Laser Vaporization	34
	2.5.3 Chemical Vapor Deposition	34
3	**Experimental**	**41**
3.1	Confocal Photoluminescence Microscope	41
	3.1.1 Overview	41
	3.1.2 The Laser System	43
	3.1.3 The Photoluminescence Microscope	44
	3.1.4 Spectrograph and Detector	48
	3.1.5 Photoluminescence Excitation (PLE) Mapping	48
	3.1.6 Photoluminescence Imaging	49

3.2	Confocal Raman Microscope	51
3.3	CVD Reactor	53

4 CVD Synthesis of Carbon Nanotubes   55
4.1	Vertically Aligned Arrays of Carbon Nanotubes	55
4.2	Suspended and Horizontally Aligned Arrays of Carbon Nanotubes	65

5 Spectroscopic Characterization of Individual Carbon Nanotubes   73
5.1	Influence of External Dielectric Screening on Optical Transition Energies		73
	5.1.1	Motivation	73
	5.1.2	Sample Preparation and Measurements	75
	5.1.3	Dielectric Screening Effect in SWNTs	83
5.2	Blinking of SWNTs at Cryogenic Temperatures		86
	5.2.1	Blinking Background	86
	5.2.2	Sample Preparation and Measurements	87
	5.2.3	Observations and Conclusions	88
5.3	Direct Observation of Deep Excitonic States in the PLE Spectra of SWNTs		97
	5.3.1	Bright and Dark Excitonic Levels	97
	5.3.2	Observations and Conclusions	98
5.4	Spectroscopy and AFM Manipulation of Individual, Ultralong Carbon Nanotubes		104
	5.4.1	Background	104
	5.4.2	Nanotubes from CO-CVD	105
	5.4.3	Nanotubes from Ethanol CVD	108
	5.4.4	Ultralong and Aligned Nanotubes from Li *et al.*	113
5.5	Individual SWNTs in Dispersion: A Method to Count Different Chiralities		125

6 Conclusion and Outlook   131

Bibliography   135

List of Publications   151

Acknowledgements   153

1 Introduction

The field of carbon nanotube research has evolved tremendously since it was triggered by the observation of carbon nanotubes on the cathode of an arc discharge apparatus used to produce fullerenes in 1991 [1]. Today, physicists, chemists, biologists, material scientists and engineers push this interdisciplinary field further in both fundamental and application-oriented directions. The large diversity of this research can be attributed to many unique properties of carbon nanotubes.

Single-walled carbon nanotubes (SWNTs) are seamless cylinders out of rolled-up strips of graphene (graphene is a monolayer of graphite). On the one hand, SWNTs can be regarded as molecular systems due to their nanometer-sized diameters, but on the other hand, their length of up to several centimeters yields an unprecedented length/diameter aspect ratio of up to 10^7 and makes them quasi-one dimensional (1D) crystals, unique in solid state physics. Additionally, the electronic band structure of carbon nanotubes strongly depends on their geometry. Changing the geometry (diameter and roll-up angle or chirality) of SWNTs, leads to either conducting or semiconducting behavior. Furthermore, the band gap of the semiconducting species can be adjusted over a wide range. No doping as in traditional semiconductors is required and all nanotube species are made purely of carbon. A wide group of semiconducting SWNTs show photoluminescence (PL) in the near-infrared spectral region, which allows for their spectroscopic investigation and characterization [2]. This spectral window also facilitates possible applications in biology and telecommunication. The atomic-scale perfection of continuous carbon nanotubes and their close relationship to graphene makes them chemically inert and is responsible for the Young's modulus of about 1 TPa for an individual SWNT [3] at a density of only $\sim 1.3\,\mathrm{g/cm^3}$.[1] Moreover, the thermal conductivity exceeds that of diamond and a single carbon nanotube can carry 1000 times more current than an imaginary copper wire of the same diameter [4].

---

[1]Steel has a typical Young's modulus of about 0.2 TPa and a density of $7.8\,\mathrm{g/cm^3}$.

1 Introduction

Owing to the aforementioned properties, applications like electrically and thermally conductive composites and fibers, field emission cathodes, field emission transistors and field emission displays, sensors, PL markers, electrodes in supercapacitors, fuel cells, Li-ion batteries and many others have been envisioned and tested [5, 6]. From the viewpoint of fundamental science, SWNTs have attracted much attention due to their quasi-1D structure. An important example is the binding energy of excitons in semiconducting SWNTs which is two orders of magnitude larger than in typical 3D semiconductors [7, 8].

One unsolved problem since the discovery of carbon nanotubes is related to the heterogeneous distribution of lengths, geometric and electronic structures and aggregation states (bundles vs. individuals) produced by current synthesis methods. An important direction of nanotube research is therefore the separation of these as-produced mixtures [6]. Apparently, heterogeneous distributions produce an average response in spectroscopic studies of SWNT ensembles. The present work focuses on another possibility, namely the investigation of SWNTs on an individual tube level. The methods used encompass micro-photoluminescence excitation (µ-PLE) spectroscopy, µ-Raman spectroscopy, PL and Raman imaging, scanning electron microscopy (SEM), atomic force microscopy (AFM) and their combination on the same nanotube. Apart from microscopic studies, this work also describes the synthesis of vertically and horizontally aligned carbon nanotube arrays via chemical vapor deposition (CVD). These as-grown SWNTs were used for several experiments including investigations of (i) the influence of a dielectric medium on the optical transition energies [9]; (ii) PL properties at low temperatures down to 4 K [10] and (iii) deep excitonic states below the optically active exciton [11]. Moreover, we characterized for the first time the PL of nanotubes grown in contact with dielectric substrates e.g. Si/SiO_2. In addition, µ-PLE spectroscopy was employed to investigate the integrity of chirality and the influence of AFM manipulation on individual SWNTs [12]. Finally, a new method to count individual, semiconducting SWNTs in water-surfactant dispersions via PL spectroscopy is presented and preliminary results are shown [13].

2 Carbon Nanotube Basics

In the following sections, a condensed theoretical background is given about the properties of carbon nanotubes. This is essential to understand their spectroscopic behavior as described in the next chapters. The background discussion starts with a characterization of the geometric structure, continues with the electronic band structure followed by the electronic density of states (DOS) of carbon nanotubes—in particular of single-walled carbon nanotubes—and concludes with their optical properties [photoluminescence (PL), absorption and Raman]. Mathematical derivations, physical arguments and notations in these sections are partially based on those given in Ref. [14]. Finally, synthesis methods of carbon nanotubes [arc discharge, pulsed laser vaporization and chemical vapor deposition (CVD)] are also discussed.

2.1 Geometric Structure

Single-walled carbon nanotubes (SWNTs) are hollow cylinders made of a single, two-dimensional (2D) layer of graphite, called graphene. All sp^2-hybridized carbon atoms are therefore surface atoms. If the tube comprises of two or more concentric hollow cylinders, it is termed a double- or multi-walled carbon nanotube (DWNT or MWNT), respectively. Typical SWNTs have diameters on the order of 1-2 nm and can nowadays be grown up to centimeters in length [15]. This aspect ratio of $\sim 10^7$ makes them unique and interesting quasi-one dimensional (1D) objects. MWNTs possess similar lengths but much larger diameters (typically 5-100 nm). The discovery of MWNTs is normally credited to S. Iijima in 1991, but publications as early as in 1952 by L. V. Radushkevich *et al.* and 1976 by A. Oberlin *et al.* showing similar high-resolution transmission electron microscopy (HRTEM) images of MWNTs should be mentioned, as well [1, 16, 17]. SWNTs were first described in 1993 independently by Ijima *et al.* and Bethune *et al.* [18, 19]. For a detailed discussion about who should be given the credit for the discovery of carbon

2 Carbon Nanotube Basics

nanotubes see Ref. [20]. SWNTs are of more fundamental interest and were mostly dealt with in this work, so the following discussion focuses on SWNTs.

The geometric structure of SWNTs is strongly related to that of graphene. Fig. 2.1 shows the hexagonal honeycomb lattice of a graphene sheet, including the definition of the primitive unit cell and the lattice vectors $\vec{a}_1$ and $\vec{a}_2$. The primitive unit cell is a rhombus containing two (highlighted) carbon atoms but only one lattice point. The lattice vectors $\vec{a}_1$ and $\vec{a}_2$ form a 60° angle and are expressed in Cartesian coordinates as:

$$\vec{a}_1 = \begin{pmatrix} \frac{\sqrt{3}a_0}{2} \\ \frac{a_0}{2} \end{pmatrix} \quad , \quad \vec{a}_2 = \begin{pmatrix} \frac{\sqrt{3}a_0}{2} \\ -\frac{a_0}{2} \end{pmatrix} \tag{2.1}$$

with $a_0 = |\vec{a}_1| = |\vec{a}_2| = \sqrt{3} a_{C-C} = 2.46\,\text{Å}$ where $a_{C-C} = 1.42\,\text{Å}$ is the carbon-carbon bond distance. The lattice vectors of graphene are now used as a basis for the so-called chiral vector $\vec{c} = n_1 \vec{a}_1 + n_2 \vec{a}_2$ along which the sheet is rolled up in such a way that $\vec{c}$ becomes the circumference of the tube. A (theoretically infinitely long) strip is cut out of the graphene lattice along two lines perpendicular to $\vec{c}$ and both are superimposed. The resulting SWNT is usually denoted a (n_1, n_2) tube. Fig. 2.1 indicates this procedure for the case of a $(6, 4)$ tube. The z-axis of a tube is perpendicular to $\vec{c}$ and points in the direction of its length. SWNTs with $n_1 = n_2$ and $(n_1, 0)$ are special cases and are called armchair and zig-zag SWNTs, respectively. The name stems from the pattern of the carbon atoms along $\vec{c}$, also indicated in Fig. 2.1 for $(6,6)$ and $(6,0)$. The chiral angle θ is measured from the zig-zag direction and can be calculated using

$$\cos\theta = \frac{\vec{a}_1 \cdot \vec{c}}{|\vec{a}_1| \cdot |\vec{c}|} = \frac{n_1 + n_2/2}{\sqrt{N}} \quad \text{with} \quad N = n_1^2 + n_1 n_2 + n_2^2 \tag{2.2}$$

For each (n_1, n_2) with $n_1 \geq n_2 \geq 0$, θ runs between 0° and 30°. For $30° \leq \theta \leq 60°$, equivalent SWNTs are formed but the chirality changes from left-handed to right-handed. (n_2, n_1) is therefore the enantiomer of a (n_1, n_2) tube (if $n_1 \neq n_2 \neq 0$). Due to the hexagonal symmetry of graphene, chiral vectors with $\theta \geq 60°$ are equal to those below 60°. Hence, the restriction to the case $n_1 \geq n_2 \geq 0$ (or $0° \leq \theta \leq 30°$) is normally sufficient. Both zig-zag ($\theta = 0°$) and armchair ($\theta = 30°$) tubes are achiral.

The circumference of a SWNT is given by the length of $\vec{c}$. Thus, the diameter can be calculated by

$$d = \frac{|\vec{c}|}{\pi} = \frac{a_0}{\pi}\sqrt{N} \tag{2.3}$$

2.1 Geometric Structure

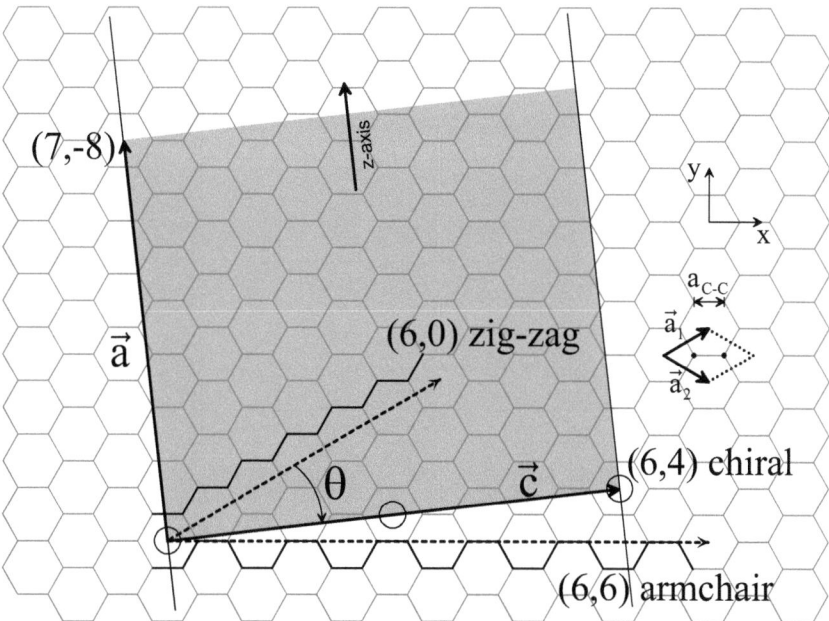

Figure 2.1: 2D graphene honeycomb lattice showing the construction of a nanotube unit cell for an exemplary (6,4) tube (grey rectangle). $\vec{a}_1$ and $\vec{a}_2$ are graphene lattice vectors used as basis for the nanotube chiral vector $\vec{c} = 6\vec{a}_1 + 4\vec{a}_2$ in circumferential direction and $\vec{a} = 7\vec{a}_1 - 8\vec{a}_2$ defining the minimum translational period in z-axis direction. The SWNT is formed by superimposing both indicated 'cutting' lines. Zig-zag and armchair direction as well as structural motifs are indicated for (6,0) and (6,6). Adapted from Ref. [14].

The size of the carbon nanotube unit cell in z-direction is determined by $\vec{a}$ (see Fig. 2.1). This vector is perpendicular to $\vec{c}$ and its length $|\vec{a}| = a$ defines the repeated translational period along the tube axis. Direction and length are given by

$$\vec{a} = \frac{2n_2 + n_1}{n\mathcal{R}}\vec{a}_1 - \frac{2n_1 + n_2}{n\mathcal{R}}\vec{a}_2 \qquad (2.4)$$

and

$$a = \frac{\sqrt{3N}}{n\mathcal{R}}a_0 \qquad (2.5)$$

where n_1 and n_2 are the chiral indices, n is the greatest common divisor of (n_1, n_2) and

2 Carbon Nanotube Basics

$\mathcal{R} = 3$ if $(n_1 - n_2)/3n$ is integer and $\mathcal{R} = 1$ otherwise. The greatest common divisor of (n_1, n_2) also corresponds to the number of graphene lattice points along $\vec{c}$, denoted by open circles in Fig. 2.1. Here, $\vec{a} = 7\vec{a}_1 - 8\vec{a}_2$ and $n = 2$ (first and last circle coincide when the sheet is rolled up). Thus, the nanotube unit cell is formed by a cylindrical surface with height a and diameter d (gray rectangle in Fig. 2.1). It contains q hexagons and, because of two carbon atoms per graphene unit cell, $2q$ carbon atoms per nanotube unit cell (n_c). q and n_c are calculated using

$$q = \frac{2N}{n\mathcal{R}} \tag{2.6}$$

$$n_c = 2q = \frac{4N}{n\mathcal{R}} \tag{2.7}$$

After having determined the unit cell of carbon nanotubes, it is important to note that the chiral indices (n_1, n_2) critically affect the electronic properties of SWNTs. Simply by going from a $(10, 10)$ to a $(10, 9)$ tube, the number of carbon atoms per unit cell increases from 40 to 1084 and the electronic properties change from metallic to semiconducting, whereas the diameter only decreases by 5% from 1.36 to 1.29 nm. The reason for this abrupt change in electronic properties will be discussed in the next section.

2.2 Electronic Band Structure

To determine the electronic band structure of SWNTs, the so-called 'zone-folding approximation' is the simplest approach. It requires the electronic band structure [or dispersion relation, $E(\vec{k})$] of a 2D graphene sheet which can be calculated using tight-binding, refined tight-binding or *ab-initio* methods. As a nanotube is a subset of a graphene sheet, the electronic band structure of a SWNT is a subset of the electronic band structure of graphene, if effects due to curvature are neglected. In the following, the Brillouin zone (BZ) of graphene and of SWNTs will be constructed in order to determine this subset.

The reciprocal lattice vectors of graphene, $\vec{k}_1$ and $\vec{k}_2$, are defined due to the translational invariance of the lattice via $\vec{a}_i \cdot \vec{k}_j = 2\pi\delta_{ij}$ with $i, j = 1, 2$ as

$$\vec{k}_1 = \begin{pmatrix} \frac{2\pi}{\sqrt{3}a_0} \\ \frac{2\pi}{a_0} \end{pmatrix} \quad , \quad \vec{k}_2 = \begin{pmatrix} \frac{2\pi}{\sqrt{3}a_0} \\ -\frac{2\pi}{a_0} \end{pmatrix} \tag{2.8}$$

Fig. 2.2 shows the orientation of $\vec{k}_1$ and $\vec{k}_2$ relative to the first Brillouin zone (BZ) of

2.2 Electronic Band Structure

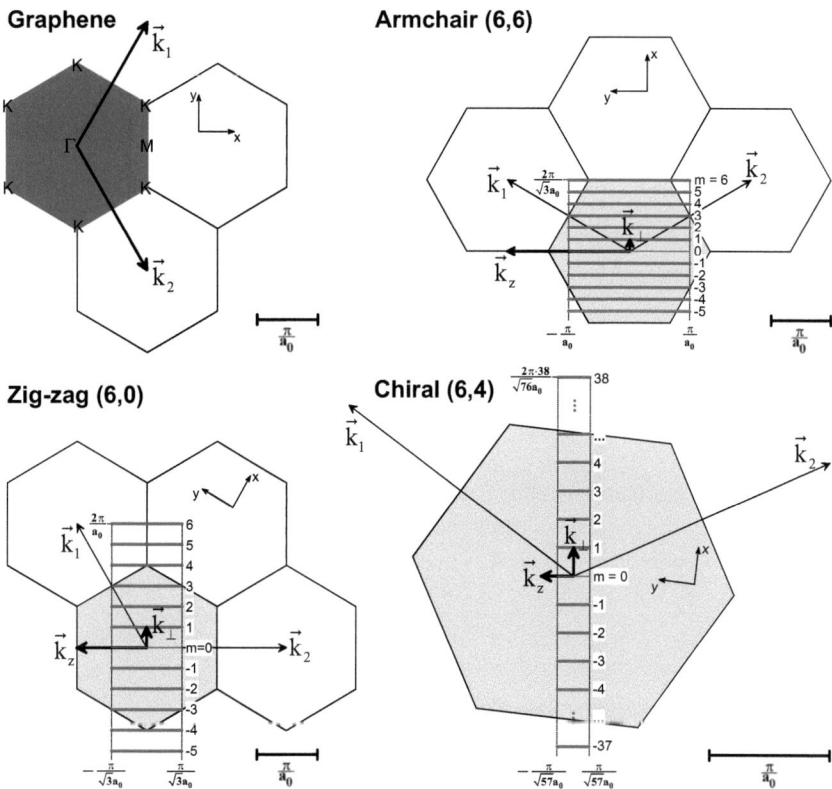

Figure 2.2: Illustration of the first BZ (in red) of graphene, a $(6,6)$ armchair, a $(6,0)$ zig-zag and a $(6,4)$ chiral carbon nanotube, constructed using the equations given in the text. Note the equal scale bars for the first 3 graphs and the doubling for the chiral SWNT. The high symmetry points Γ, K and M are indicated for graphene as well as the dimensions of the BZs of the three nanotubes. The BZ of the chiral SWNT is too long to be fully depicted in the '$\vec{k}_\perp$-extended representation' [21]. It consists of $q = 76$ parallel lines.

graphene which is depicted in red. Together, they constitute an angle of $120°$. The diagram also defines the positions of the high symmetry points Γ, K and M. Γ is the center of the BZ at $\vec{k} = 0$. K and M are located at the vertices and in the middle of

2 Carbon Nanotube Basics

each side of the BZ, respectively.

The reciprocal lattice vectors of a SWNT are calculated as for graphene using $a_i \cdot k_j = 2\pi\delta_{ij}$, with a_i now being $\vec{a}$ and $\vec{c}$, respectively. Reciprocal lattice vectors are termed $\vec{k}_z$ for wave movement along the z-axis and $\vec{k}_\perp$ for circumferential direction. The calculation yields:

$$\vec{k}_z = \left(\frac{n_2}{q}\right)\vec{k}_1 - \left(\frac{n_1}{q}\right)\vec{k}_2 \qquad (2.9)$$

$$\vec{k}_\perp = \left(\frac{2n_1 + n_2}{qn\mathcal{R}}\right)\vec{k}_1 + \left(\frac{2n_2 + n_1}{qn\mathcal{R}}\right)\vec{k}_2 \qquad (2.10)$$

For an infinitely long carbon nanotube, $\vec{k}$-vectors along $\vec{k}_z$ are continuous, i.e. they can take any value. The modulus of $\vec{k}_z$ corresponds to the translational period a via:

$$|\vec{k}_z| = k_z = \frac{2\pi}{a} \qquad (2.11)$$

However, there are boundary conditions for $\vec{k}_\perp$, because in circumferential direction a wave function can interfere with itself after one cycle. Hence, $\vec{k}_\perp$ becomes quantized. Along the circumference, only integer multiples of the wavelength λ yield stationary results. All other wavelengths will vanish by interference:

$$m \cdot \lambda = |\vec{c}| = \pi \cdot d \quad \Rightarrow \quad |\vec{k}_{\perp,m}| = k_{\perp,m} = \frac{2\pi}{\lambda} = \frac{2\pi}{|\vec{c}|} \cdot m, \qquad (2.12)$$

where m is an integer which takes values $-q/2 + 1, ..., 0, 1, ..., q/2$ [1]. The first BZ of a SWNT therefore consists of q lines parallel to the z-axis separated by $|\vec{k}_\perp| = k_\perp = 2/d$ and of width $2\pi/a$. Fig. 2.2 shows the first BZ in red for all SWNTs indicated in Fig. 2.1 [armchair $(6,6)$, zig-zag $(6,0)$ and chiral $(6,4)$], always relative to the first BZ of graphene, denoted in gray. For all three carbon nanotubes, $\vec{k}_\perp$ is chosen to point up and all graphs (including the diagram for graphene) are drawn to scale. For the $(6,4)$ SWNT, the big translational period a yields a very narrow first BZ. In addition, it consists of $q = 76$ parallel lines. Therefore, the scale was doubled relative to the other three and only the central part of the first BZ is shown. In general, the number of allowed k-lines increases (decreases) with increasing (decreasing) diameter, whereas their distance from each other decreases (increases).

---

[1] This derivation uses the same arguments as for the rigid rotor. Thus, m also corresponds to the quantum number of the electron angular momentum in z-axis direction which will be important for the selection rules of optical transitions in carbon nanotubes in Sec. 2.4

2.2 Electronic Band Structure

The electronic band structure of SWNTs can now be calculated by using the dispersion relation of graphene, $E(\vec{k})$, and restricting $\vec{k}$ to vectors which belong to the BZ of a carbon nanotube. Because the zone-folded line segments 'cut' the relevant energies out of the band structure of graphene, they are also called 'cutting lines' [21].

The electronic band structure of graphene is shown in Fig. 2.3. Figure 2.3(a) was calculated using the nearest-neighbor tight-binding approach with carbon $2s$ and $2p$ atomic orbitals as basis (nearest-neighbor exchange integral $\gamma_0 = -3.033$ eV and nearest-neighbor overlap integral $s_0 = 0.129$ eV for the π bands [22]). The Fermi level is set to zero. Note that the π valence and π^* conduction band touch at the K points. Thus, graphene is a semi-metal (also called a zero gap semiconductor) as the Fermi surface consists of only 6 points, i.e. the 6 K points of the BZ. To explain optical properties of graphene and of carbon nanotubes, it is often sufficient to consider the π bands around the K points only, as the smallest gap between the σ bonding and the σ^* antibonding bands is ~ 6 eV. A 3D plot of the π valence and π^* conduction band is depicted in Fig. 2.3(b). Here, $\gamma_0 = -2.79$ and s_0 is approximated to zero making the 2 bands symmetric. This is a good approximation for energies around the Fermi level and is mostly used to derive zone-folding results for carbon nanotubes. The value of γ_0 is normally derived from a fit to experimental results and varies between 2.5 eV and 3 eV, depending on the specific experimental method. Figure 2.3(c) shows an enlarged contour plot of the π^* conduction band around the K point. Note the trigonal symmetry of the contour lines extending towards the M points. In Sec. 2.3 the purely circular contour shown in Fig. 2.3(d) will be used to calculate the density of states (DOS) in a zero order approximation. The differences caused by this approximation are generally called 'trigonal warping effects'.

The explanation why SWNTs are semiconducting or metallic depending on their chiral indices (n_1, n_2) is now straightforward in the zone-folding approximation. If the K point of the graphene BZ is part of the nanotube BZ it will be metallic and semiconducting otherwise. The BZs of the armchair $(6,6)$ and the zig-zag $(6,0)$ tube in Fig. 2.2 intersect the K point of graphene and therefore both tubes are metallic. The $(6,4)$ tube misses the K point of the first BZ (and also of the adjacent BZs, not shown) and exhibits a semiconducting character. In general, the scalar product of two vectors belonging to two reciprocal lattices is a multiple of 2π. This property is due to the translational invariance of a lattice. If this correlation is applied to the K point of graphene at $\vec{k}_K = \frac{1}{3}(2\vec{k}_1 + \vec{k}_2)$

2 Carbon Nanotube Basics

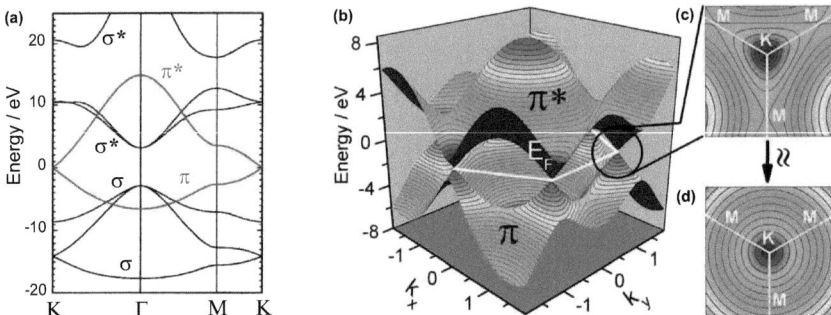

Figure 2.3: Different representations of the electron energy dispersion relation for graphene. **(a)** shows the bands calculated in the tight-binding approximation along the path K-Γ-M-K (γ_0 = -3.033 eV, s_0 = 0.129 eV). π valence and π^* conduction band touch at the K points and are highlighted in red. Modified from Ref. [22]. **(b)** is a 3D diagram of only π and π^* bands with s_0 approximated to zero, making the two bands symmetric (γ_0 = -2.79 eV). **(c)** shows an enlarged contour plot of the π^* conduction band around the K point, exhibiting trigonal symmetry. Close to K, the circular contour depicted in **(d)** is often used for a first approximation.

and the nanotube chiral vector $\vec{c}$, one gets:

$$\vec{k}_K \cdot \vec{c} = 2\pi p = \frac{1}{3}\left(2\vec{k}_1 + \vec{k}_2\right)(n_1\vec{a}_1 + n_2\vec{a}_2) = \frac{2\pi}{3}(2n_1 + n_2) \quad \text{with} \quad p \in \mathbb{Z} \quad (2.13)$$

using $\vec{a}_i \cdot \vec{k}_j = 2\pi\delta_{ij}$ with $i,j = 1,2$. Transformation yields

$$3p = 2n_1 + n_2 \quad (2.14)$$

which states that a tube is metallic if $2n_1 + n_2$ is a multiple of three, or, equivalently, if $n_1 - n_2$ is a multiple of three [the latter is the result when taking the K point at $\vec{k}_K = \frac{1}{3}\left(\vec{k}_1 - \vec{k}_2\right)$]. Hence, 1/3 of all possible nanotubes are metallic and 2/3 are semiconducting. All armchair tubes are metallic. The exact position of the K point between two cutting lines for a semiconducting nanotube can be calculated by dividing the projection of $\vec{k}_K$ on $\vec{k}_\perp$ by $|\vec{k}_\perp|$:

$$\frac{\vec{k}_K \cdot \vec{k}_\perp}{|\vec{k}_\perp|^2} = \frac{2n_1 + n_2}{3} \quad (2.15)$$

or, again equivalently, $(n_1 - n_2)/3$ for the other K point. If $(2n_1 + n_2)/3 = p$ is integer, the pth cutting line (counted from the Γ point) hits the K point and the tube is metallic.

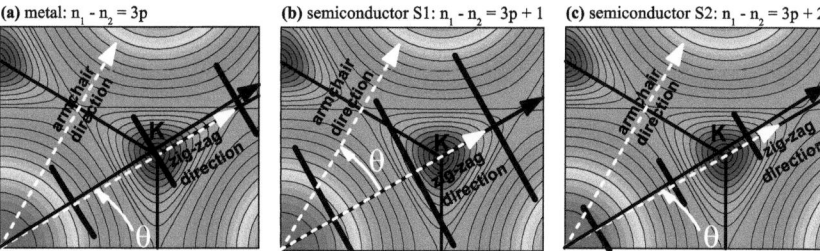

Figure 2.4: Three different configurations of nanotube cutting lines in the vicinity of the K point drawn on top of the energy contour for the π^* conduction band of graphene. **(a)** Metallic nanotube, $(n_1 - n_2)$ mod $3 = 0$ with one of the cutting lines intersecting the K point. **(b)** and **(c)** Semiconducting nanotubes from the S1 and S2 family with $(n_1 - n_2)$ mod $3 = 1$ and $(n_1 - n_2)$ mod $3 = 2$, respectively. Note the different energy contours for the cutting lines closest to K in (b) and (c), respectively. All three SWNTs are zig-zag or close to zig-zag tubes and similar in diameter, e.g. (10,1), (10,0) and (9,1).

If $(2n_1 + n_2)$ is not divisible by three i.e. if $(2n_1 + n_2)$ mod $3 = 1$ or 2, then the K point is either at $1/3$ or $2/3$ of the distance between two adjacent cutting lines, giving rise to two different, so-called 'families' of semiconducting nanotubes, S1 and S2. All three cases are shown in Fig. 2.4 on top of the energy contour of the π^* conduction band of graphene. If this contour were circularly symmetric around K, S1 and S2 SWNTs would be equivalent. But due to trigonal warping, the cutting line closest to K for S1 samples a different energy contour compared to the same line for S2. The consequences of trigonal warping will also be discussed in the next section.

2.3 Electronic Density of States

In the discussion of electronic and optical properties of SWNTs, the density of (electronic) states (DOS) is an essential quantity. It is calculated from the electronic band structure and therefore depends dramatically on the dimensionality of the system. For 3D solids, the DOS typically rises proportionally to the square root of the energy. In the 2D case, it exhibits a step-like function. In 1D systems like nanowires and quasi-1D systems like nanotubes, a $1/\sqrt{E}$-behavior is expected. Finally, the DOS is normally a δ-function in 0D systems (like molecules or quantum dots). The density of states, $n(E)$

2 Carbon Nanotube Basics

for q 1D bands, (i.e. cutting lines with index $m = -q/2 + 1...q/2$), $E_m(k_z)$, can be expressed as [23]:

$$n(E) = \frac{\partial N(E)}{\partial E} = \frac{2}{q|\vec{k}_z|} \sum_m \sum_i \int \left|\frac{dE_m(k_z)}{dk_z}\right|^{-1} \delta(k_z - k_{i,m}) \, dk_z \qquad (2.16)$$

where $k_{i,m}$ is defined by $E - E_m(k_{i,m}) = 0$ i.e. the points where, for a given energy E, a horizontal line $E = const.$ intersects with $E_m(k_z)$. Integration over the Dirac δ-function takes the inverse derivatives at these points and the summation over i adds them up. The sum over m includes the case of $E = const.$ hitting more than one band. The factor 2 accounts for the two spin states and $q|\vec{k}_z|$ is the total length of the nanotube BZ.[2] $N(E)$ is the total number of electronic states per graphene unit cell below a given energy E. Therefore, $n(E)$ corresponds to the number of states per unit energy and graphene unit cell or the DOS per every two carbon atoms.

The simplest approximation of the graphene band structure close to the Fermi level (K point) is given by a linear dispersion relation:

$$E(\vec{k}) = \pm\frac{\sqrt{3}}{2} a_0 \gamma_0 |\vec{k} - \vec{k}_K| \quad \text{with} \quad |\vec{k} - \vec{k}_K| = \sqrt{\Delta k_{\perp,m}^2 + \Delta k_z^2} \qquad (2.17)$$

where $\Delta k_{\perp,m}$ is the quantized component along $\vec{k}_\perp$ and Δk_z is the component along $\vec{k}_z$. The upper part of this double cone is depicted in Fig. 2.3(d) as a contour diagram. In order to find the quantization condition for $\Delta k_{\perp,m}$, the projection of $\vec{k} - \vec{k}_K$ on $\vec{k}_\perp$ is calculated:

$$\Delta k_{\perp,m} = (\vec{k} - \vec{k}_K) \frac{\vec{k}_\perp}{|\vec{k}_\perp|} = \frac{2}{3d}(3m - n_1 + n_2) \qquad (2.18)$$

This quantization forms the cutting lines on the double cone as shown in Fig. 2.5(a). $E(\vec{k})$ of Eq. 2.17 now simplifies to $E_m(\Delta k_z)$ and the inverse of the derivative in Eq. 2.16 is

$$\left|\frac{dE_m(k_z)}{dk_z}\right|^{-1} = \left|\frac{dE_m(\Delta k_z)}{d(\Delta k_z)}\right|^{-1} = \frac{2}{\sqrt{3} a_0 \gamma_0} \left|\frac{d\sqrt{\Delta k_{\perp,m}^2 + \Delta k_z^2}}{d(\Delta k_z)}\right|^{-1} \qquad (2.19)$$

$$= \frac{2}{\sqrt{3} a_0 \gamma_0} \frac{|E_m(k_z)|}{\sqrt{E_m(k_z)^2 - \epsilon_m^2}} \qquad (2.20)$$

---

[2] Note that in the integral, k_z is a variable, ranging between $-\pi/a$ and π/a, whereas $|\vec{k}_z|$ denotes the length of the nanotube BZ in z-direction.

2.3 Electronic Density of States

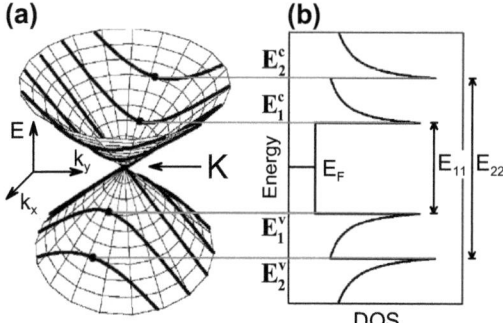

Figure 2.5: (a) Band structure of the π and π^* bands around the K point of graphene, approximated as two cones. The boundary condition for a metallic SWNT leads to q cutting lines, one of them intersecting at the K point. (b) After calculation of the DOS according to Eq. 2.16, the points of zero slope correspond to van-Hove singularities of the valence (E_i^v) and conduction bands (E_i^c), respectively. i increases away from the Fermi level. Note the non-zero DOS around the Fermi level for a metallic tube. E_{11} and E_{22} are transition energies for the first and second optical excitation for light polarized parallel to the nanotube axis, respectively. Modified from Ref. [21].

with

$$\epsilon_m = \frac{\sqrt{3}}{2} a_0 \gamma_0 \Delta k_{\perp,m} = (3m - n_1 + n_2) \frac{a_0 \gamma_0}{\sqrt{3}d} \qquad (2.21)$$

At the energies $\epsilon_m \neq 0$, the derivative becomes zero and its inverse, and therefore the DOS, diverges. These points of zero slope are called van-Hove singularities and are characteristic for 1D systems. In Fig. 2.5(a), they are depicted as points on the cutting lines. When inserting Eq. 2.20 into Eq. 2.16, the integration can be performed after calculating the k_i. Due to the symmetry of one cone, every nanotube band that is intersected by a line $E = const.$ is intersected twice (as long as $|E| \neq |\epsilon_m|$), with equal modulus of the slope. Thus, the summation over i can be resolved to a factor of 2:

$$n(E) = \frac{4}{q|\vec{k}_z|} \sum_{m=-q/2+1}^{q/2} \frac{2}{\sqrt{3}a_0\gamma_0} g(E, \epsilon_m) \qquad (2.22)$$

$$= \frac{2a_0}{\pi^2 d \gamma_0} \sum_{m=-q/2+1}^{q/2} g(E, \epsilon_m) \qquad (2.23)$$

13

2 Carbon Nanotube Basics

with

$$g(E, \epsilon_m) = \begin{cases} \frac{|E|}{\sqrt{E^2-\epsilon_m^2}} & |E| > |\epsilon_m| \\ 0 & |E| < |\epsilon_m| \end{cases} \quad (2.24)$$

Fig. 2.5 demonstrates the key points of the previous derivation for the case of a metallic SWNT and the labeling of the van-Hove singularities. For metallic nanotubes, the DOS around the Fermi energy is finite whereas for semiconducting ones, E_F is located in the middle of the band gap. According to Eq. 2.21, the energies of the van-Hove singularities are given by $\epsilon_m = \pm j a_0 \gamma_0 / \sqrt{3} d$ with $j = 3, 6, 9, \ldots$ for metallic and $j = 1, 2, 4, 5, 7, 8\ldots$ for S1 and S2 semiconducting SWNTs. $\epsilon_m = 0$ corresponds to the Fermi level. The energy differences E_{ii}, $i = 1, 2, \ldots$ between equal pairs of van-Hove singularities in the valence and conduction band are experimentally accessible e. g. using absorption or photoluminescence (PL) spectroscopy. Calculation of E_{11} and E_{22} yields:

$$\begin{aligned} &\text{For metallic SWNTs:} && E_{11}^M = 6a_0\gamma_0/\sqrt{3}d \,, && E_{22}^M = 12a_0\gamma_0/\sqrt{3}d \\ &\text{For semiconducting SWNTs:} && E_{11}^S = 2a_0\gamma_0/\sqrt{3}d \,, && E_{22}^S = 4a_0\gamma_0/\sqrt{3}d \end{aligned} \quad (2.25)$$

Thus, in a zero order approximation, theory predicts the transition energies to be inversely proportional to the diameter of the nanotube. A diagram plotting the transition energies E_{ii} vs. tube diameter is generally called 'Kataura plot' after Kataura et al. [24] who first used this graphical representation. Fig. 2.6(a) shows a theoretically calculated Kataura plot using tight-binding with $\gamma_0 = -2.9\,\text{eV}$ and $s = 0$, but including trigonal warping. Each dot in the figure corresponds to one transition energy $E_{ii}(d)$ of one pair of chiral indices, (n_1, n_2). The relations given in Eq. 2.25 form the central line of a band, within which the dots are located. They do not collapse on a single line due to the trigonal warping effect. The spreading increases for decreasing diameter and increasing index i. Both can be understood in the context that the approximation of a linear dispersion around the K point of graphene is only valid close to K. A decreasing diameter leads to a larger spacing between adjacent cutting lines (the spacing is equal to $2/d$) and an increasing i corresponds to cutting lines further away from the Fermi level. Within one $E_{ii}(d)$ band, metallic armchair tubes or semiconducting tubes close to armchair fulfill the $1/d$-behavior best and are located close to the middle of the band, whereas zig-zag SWNTs deviate strongest and form the upper and lower bound thereof. The direction of the deviation depends on $(n_1 - n_2) \bmod 3$ which is seen best in Fig. 2.6(b). Here, PL data of semiconducting SWNTs are plotted for the typical diameter range of a sample.

2.3 Electronic Density of States

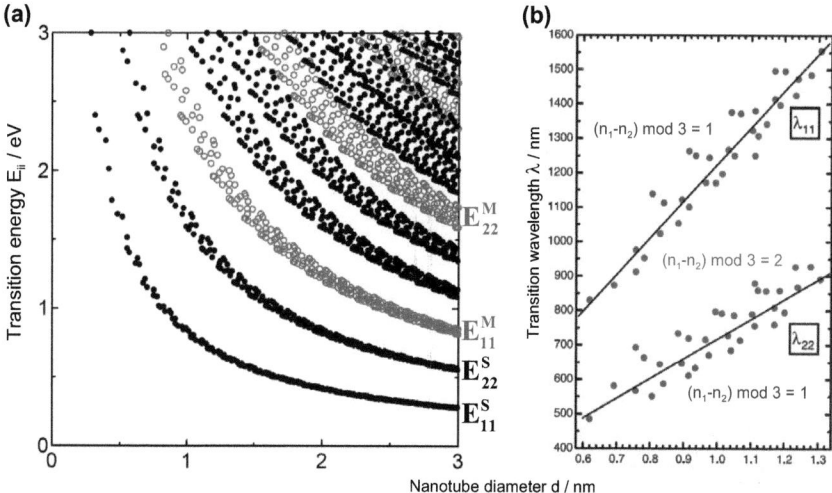

Figure 2.6: (a) calculated tight-binding Kataura plot with $\gamma_0 = -2.9\,\text{eV}$ and $s = 0$. Optical transitions symmetric to the Fermi level are black and red for semiconducting and metallic SWNTs, correspondingly. The approximate $1/d$-curves from Eq. 2.25 run through the middle of the bands. Modified from Ref. [25] (b) Experimental Kataura plot from PL spectroscopy for λ_{ii}^S, $i = 1, 2$. S1 and S2 tubes are always on opposite sides of the $(\lambda \propto d)$-median and the sides are alternating with i. Modified from Ref. [2].

The ordinate in this experimental Kataura plot is the transition wavelength instead of the energy, forming a linear $\lambda_{ii}(d)$ relation, depicted as two straight black lines. Qualitatively, the trigonal warping effect apparent in the experimental data shows the same trend as the calculation: it tends to 'fan out' the linear relationship. Tubes close to armchair [≘ small $(n_1 - n_2)$ ≘ θ close to 30°] sit close to the median line, and SWNTs close to zig-zag [≘ large $(n_1 - n_2)$ ≘ θ close to 0°] are further away. Additionally, λ_{11} (λ_{22}) is bigger (smaller) for S1 tubes and smaller (bigger) for tubes belonging to the S2 family. The position below or above the median line for a certain nanotube alternates with the band index i. This behavior can also be deduced from Fig. 2.4 when considering the orientation of the cutting lines of armchair and zig-zag tubes relative to the orientation of the energy contour around the K point.

The approximations of simple tight-binding, zone folding, linear dispersion around K

2 Carbon Nanotube Basics

and the inclusion of trigonal warping presented in the last two sections are sufficient (and computationally much cheaper than *ab initio* methods) for a qualitative understanding of the properties of carbon nanotubes. So far, the effects of curvature as well as Coulomb interaction between electrons were ignored. The results for long and narrow stripes of graphene would have been essentially the same as for carbon nanotubes, and tight-binding explicitly excludes electron-electron interaction. The effect of curvature, beside a band-shift, is an opening of secondary energy gaps in the DOS of metallic nanotubes around the Fermi level. Since this thesis mainly deals with semiconducting SWNTs and since the gaps are quite small (on the order of kT), this effect will not be discussed further. However, Coulomb interactions have a great influence on optical properties of SWNTs as will be explained in the next section.

2.4 Optical Properties

2.4.1 Overview

The most important optical methods used for the investigation of carbon nanotubes are (resonance) Raman scattering, absorption and photoluminescence (PL) spectroscopy. Raman spectroscopy was the first to be applied to bulk amounts of SWNTs in 1994 as this method requires little or no sample preparation [26, 27]. Optical absorption spectroscopy of thin films of SWNTs was introduced in 1999 by Kataura *et al.* [24] and in 2002, Bachilo *et al.* first measured photoluminescence of SWNTs [2]. The reason why the PL of SWNTs was first shown almost 10 years after the discovery of SWNTs was, on the one hand, due to the low purity of nanotube material in early synthesis methods (i.e. arc discharge methods [28]) and on the other, the requirement to isolate SWNTs from environmental interactions. Carbon nanotubes tend to form hexagonal-packed bundles and ropes during the growth process because of high van der Waals binding energies of typically $300\,\text{meV}/\text{Å}$ for a tube inside a bundle [29]. Intermolecular tube-tube interaction average out the unique characteristics of individual SWNTs like chirality dependent band gaps and sharp van Hove singularities. The need to break up the bundles led to the use of surfactants, which, with the help of strong ultrasonic treatment, are able to individualize the tubes and enclose them in a micelle, thus forming a metastable dispersion and preventing reagglomeration. Bundles enclosed in a micelle exhibit higher densities than individual ones and therefore allow a separation via ultracentrifugation. 'Standard' surfactants are sodium dodecyl sulfate (SDS), sodium dodecyl benzylsulfonate (SDBS), sodium cholate (SC) and DNA/RNA fragments. The solvent used is normally D_2O because heavy water is almost transparent for wavelengths up to 1800 nm [30] and the band gap of typical semiconducting SWNTs extends from ~ 800 nm to the near infrared.

It is assumed that every SWNT synthesis method generates a more or less broad Gaussian distribution of diameters (and lengths). Although it is highly desirable to find a chirality selective production process, this has not yet been achieved and is part of ongoing research (see e.g. Ref. [31]). Dispersing the tubes with a surfactant and using ultracentrifugation can introduce selectivity e.g. according to diameter [32], electronic type [33] and even chirality [34, 35] if the parameters are chosen accordingly. In general, however, it is expected that the SWNT diameter distribution pertaining to a D_2O/SDS/SWNT dispersion is only slightly different from that of the raw material. Therefore, spectroscopic techniques applied to bulk amounts of such dispersions will typically probe an ensemble of SWNTs characteristic for the as-prepared material. Their features will overlap and complicate the assignment to certain chiralities. The investigation of single SWNTs, however, avoids the disadvantages of ensemble averaging and the overlap of spectral features from tubes with different chiralities. Because bulk nanotube samples comprise of such a heterogeneous mixture of lengths, chiralities, bundles and metallic/semiconducting types, single molecule spectroscopy techniques like micro (μ)-Raman, -PL and -absorption have been applied to individual SWNTs or small ensembles thereof. This work deals with μ-Raman and μ-PL(E) spectroscopy[3], but there are other, more specialized techniques on the single tube level like Rayleigh scattering spectroscopy [36–38], photocurrent [39] and electroluminescence spectroscopy [40, 41].

Either of the three standard spectroscopic techniques (Raman, absorption and PL), for individual tubes or ensembles, requires the absorption/emission of a photon in the optical frequency range by a SWNT. The selection rules for such a process will be explained in the following subsection.

2.4.2 Selection Rules

An optical photon absorbed by a SWNT will excite an electron from the valence to the conduction band and leave a hole behind. Due to the high DOS, optical absorption will be dominated by transitions between different van Hove singularities. These transitions

---

[3]PLE stands for 'photoluminescence excitation', see subsection 2.4.3

2 Carbon Nanotube Basics

cannot change the wave vector k_z of the scattered electron because the photon wavelength ($\lambda \approx 1\,\mu\mathrm{m}$) is much larger than the length of the carbon nanotube unit cell ($a \approx 10\,\mathrm{nm}$):

$$k_{ph} = \frac{2\pi}{\lambda} \ll k_z = \frac{2\pi}{a} \qquad (2.26)$$

Thus, the transition can be considered vertical, i.e. $\Delta k_z \approx 0$ in a band structure diagram $E(k_z)$ which is, apart from the DOS picture, generally used to visualize optical transitions (see Fig. 2.7).

Apart from vertical transitions, the selection rules depend on the polarization of the electric field vector relative to the nanotube axis. Closely related to this is the band index m which, except for running from $-q/2 + 1$ to $q/2$, is also the electron angular momentum quantum number in z-axis direction. How it changes upon absorption of a photon can be decided when considering the conservation of angular momentum. As bosons, photons possess a spin of $\pm \hbar$ (left and right circularly polarized light) along the propagation axis. If this axis is parallel to the z-axis of the SWNT, Δm changes by $+1$ (for left handedness) or -1 (for right handedness). This is equivalent to light linearly polarized (which can be regarded as a linear combination of equal amplitudes of left and right circularly polarized light) perpendicular to the z-axis. If the axis of propagation is perpendicular to the nanotube z-axis, m cannot change for circularly polarized light. This corresponds to the situation of light linearly polarized along the z-axis. In this thesis and in most other experiments, linear polarization is used, therefore:

$$\begin{aligned} \Delta m &= 0 \quad \text{for} \quad \vec{E} \parallel z \quad \text{and} \\ \Delta m &= \pm 1 \quad \text{for} \quad \vec{E} \perp z \end{aligned} \qquad (2.27)$$

Transitions with $\vec{E} \parallel z$ are also called longitudinal, symmetric or parallel, and transverse, asymmetric or cross-polarized for $\vec{E} \perp z$. The dipole selection rules from Eq. 2.27 still mostly apply after a more rigorous treatment using symmetry arguments [42–44] and transition matrix element calculations [45], with the exception of some cross-polarized transitions weakened/forbidden due to symmetry reasons [14]. The latter are of minor importance for this thesis which deals mostly with parallel transitions.

For an assessment of the absorption cross section of a SWNT for different linear polarizations, the 'depolarization effect' or 'antenna effect' plays a key role, as well [46–48]. The underlying argument for this phenomenon is that for the calculation of optical transition intensities in the dipole approximation, not the external electrical field

2.4 Optical Properties

strength $\vec{E}_{ext}$ of the incident light but the local electric field, $\vec{E}_{loc}$ is important. $\vec{E}_{loc}$ can be expressed as the sum of $\vec{E}_{ext}$ and a depolarization field $\vec{E}_{dp}$. $\vec{E}_{dp}$ is due to induced charges and aligns antiparallel to $\vec{E}_{ext}$. These induced charges form dipole moments which align parallel to $\vec{E}_{ext}$ [convention assigns opposite directions to the electric field (from + to −) and the dipole moment (from − to +)]. Thus, the electric polarization density $\vec{P}$ of the material ($\vec{P}$ represents the dipole moment per volume) is parallel to $\vec{E}_{ext}$.

$$\vec{E}_{loc} = \vec{E}_{ext} + \vec{E}_{dp} = \vec{E}_{ext} - \frac{N\vec{P}}{\epsilon_0} \quad (2.28)$$

with ϵ_0 being the permittivity of vacuum and N denoting the depolarization factor, which depends on the geometry of the material and its orientation relative to the electric field [49]. In a static approximation, no dipole moment is induced in an infinitely long cylinder for $\vec{E} \parallel z$ and $\vec{E}_{ext} = \vec{E}_{loc}$, $N = 0$. However, if $\vec{E} \perp z$, charges are induced on the cylinder walls that screen $\vec{E}_{ext}$ and $\vec{E}_{loc}$ is reduced, with $N = 1/2$ [49]. This geometric effect is due to the high aspect ratio of carbon nanotubes and leads to a strong suppression of transverse transitions (with $\Delta m = \pm 1$) in randomly aligned nanotubes.

Nevertheless, Maruyama et al. have attempted to measure transverse Raman spectra on vertically aligned carbon nanotube films or 'forests' [50] as well as carrying out polarized PL spectroscopy in solution and in gelatin matrix [51, 52]. Some of the observed weak features could be attributed to E_{12} and E_{21} transitions which should coincide halfway between E_{11} and E_{22}, using the linear dispersion approximation around the graphene K point (see Sec. 2.3). Asymmetric transitions with $\Delta m = \pm 1$ are transitions between adjacent cutting lines. Thus, momentum conservation requires the angular momentum of the photon along z, $\pm\hbar$, to increase (decrease) the electron wave vector in circumference direction by $k_\perp = \pm 2/d$:

$$L_{z,ph} = \pm\hbar = r_{tube} \cdot p_{circ} = \frac{d}{2} \cdot \hbar \cdot k \quad \Leftrightarrow \quad k = k_\perp = \pm\frac{2}{d} \quad (2.29)$$

Figure 2.7(a) shows different optically allowed longitudinal and transverse transitions in a simplified band structure diagram $E(k_z)$ for a typical semiconducting carbon nanotube with a band gap (E_g) of ∼1 eV. Transitions 1-3 are longitudinal transitions with 1 being the most probable, because it takes place between two van-Hove singularities. Transition 1 and 2 are absorption processes whereas transition 3 corresponds to the emission of a photon. Transition 4 depicts a cross-polarized absorption (E_{21}): k_z remains constant (the photon wave vector is negligible) but $\vec{k}_{\perp,m}$ changes by $\pm 2/d$ (transfer of angular

2 Carbon Nanotube Basics

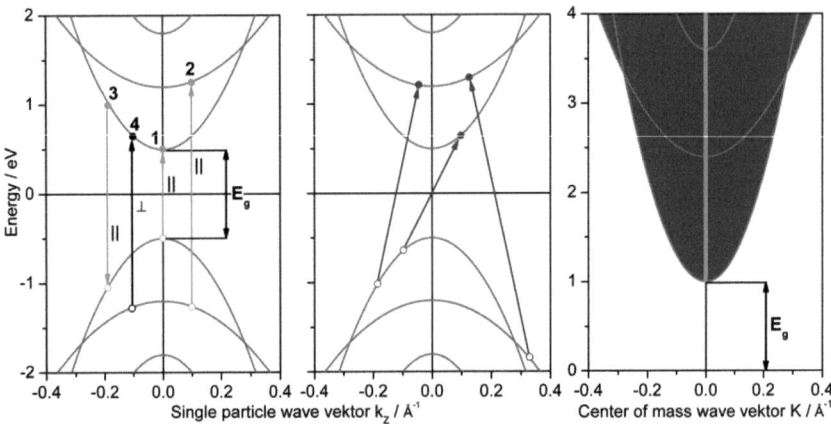

Figure 2.7: Exemplary transitions in a simplified band structure diagram for a semiconducting SWNT. E_g denotes the band gap. **(a)** Vertical optical absorption and emission processes. Transitions 1-3 are allowed for $\vec{E} \parallel z$ and 4 for $\vec{E} \perp z$. **(b)** Optically forbidden, parallel transitions. **(c)** Useful diagram for parallel transitions using the center of mass wave vector K instead of the single-particle wave vector k_z. Note that all transitions with $K = 0$ [green arrows in **(a)**] and all (parallel) transitions with $K \neq 0$ [blue arrows in **(b)**] can be represented by the green line and the blue area, respectively. Modified from Ref. [53].

momentum, + or − depends on the actual m indices of the bands involved). Since there are two quasi-particles, an electron and a hole and two kinds of transitions for parallel and cross-polarized light, it is useful to define a center of mass wave vector $\vec{K}$ which is a 2D vector in the unfolded BZ of a SWNT:

$$\vec{K} = \frac{\vec{k}_c - \vec{k}_v}{2} \qquad (2.30)$$

where $\vec{k}_c$ and $\vec{k}_v$ are the wave vectors of electron and hole in the **c**onduction and **v**alence band, respectively. The origin of these vectors is the Γ point of the graphene BZ. For transitions 1-3 in Fig. 2.7(a), $\vec{K} = 0$. The transverse transition 4 has $\vec{K} = \pm \vec{k}_\perp/2$. Figure 2.7(b) depicts parallel transitions with $\vec{K} \neq 0$. Such transitions are not allowed optically. However, they become possible if phonon-assisted transitions are considered. The diagram in Fig. 2.7(c) is a convenient representation of the center of mass energy vs. the center of mass wave vector for parallel transitions. All transitions indicated by

2.4 Optical Properties

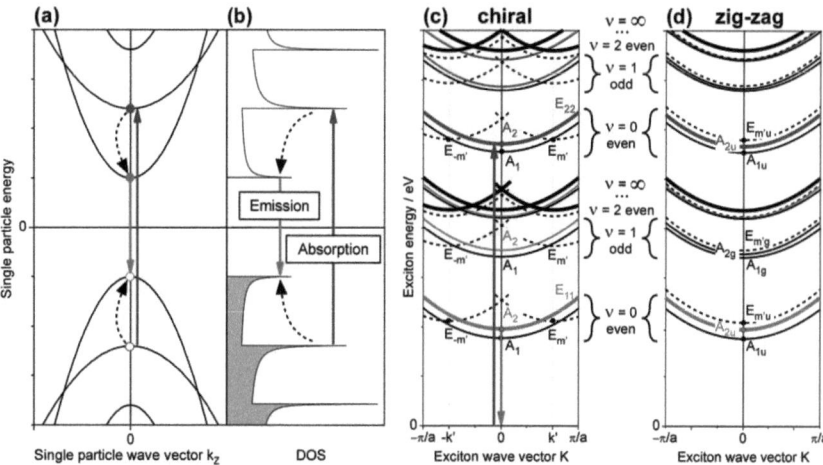

Figure 2.8: Schematic diagrams explaining PL and participating states in semiconducting SWNTs. **(a)** and **(b)** Electron and hole in an oversimplified single-particle dispersion relation and DOS diagram. '0' corresponds to the Fermi level. **(c)** and **(d)** more accurate description of singlet excitonic states and their irreducible representations in the wave vector group notation for chiral and zig-zag tubes. A similar diagram can be constructed for metallic armchair SWNTs. Colored (blue and red) bands show optically allowed, i.e. bright states. All other states are dark. They are grouped according to the exciton envelope function $F_\nu(z_e - z_h)$ which is either even ($\nu = 0, 2, 4...$) or odd ($\nu = 1, 3, 5...$) upon $z \to -z$ operations. ν labels the levels in the 1D hydrogenic series. Adapted from Ref. [54–57].

a green arrow in Fig. 2.7(a) collapse onto the green line at $\vec{K} = 0$ and all transitions in (b) with $\vec{K} \neq 0$ are within the blue area. Similar diagrams can be constructed for cross-polarized transitions.

2.4.3 Photoluminescence and Absorption Spectroscopy

PL and absorption spectra can now be interpreted in terms of the selection rules given in the last subsection. Figure 2.8(a) and (b) show the processes of absorption (blue arrow), relaxation to the band gap (dotted arrow) and the emission of PL (red arrow) in a band structure and DOS diagram for a semiconducting SWNT. Typically, transitions

21

involving the second van Hove singularity are used to excite PL but higher excitations (E_{ii} with $i > 2$) are also possible. In order to find all semiconducting chiralities present in a sample, photoluminescence excitation (PLE) spectroscopy is used, where PL emission spectra are recorded while the excitation wavelength is changed. This builds up a 3D 'map' showing PL intensity vs. emission and excitation wavelength. Every spot in a PLE map of SWNTs represents a single (n_1, n_2) species. For an example of such a 'map' which is often depicted as a contour plot, see page 76.

When the energy of an incoming parallel polarized photon is equal to E_{ii}, it is absorbed, creating an electron in the conduction band and a hole in the valence band. Both quickly (on the order of 100 fs [58]) relax to either the band gap of a semiconducting tube or to the Fermi surface of a metallic SWNT. The excess energy is emitted via phonons. In a metallic carbon nanotube, electron and hole always recombine non-radiatively whereas in semiconducting species, a radiative decay via photon emission is possible, leading to PL. From the absence of strong PL in SWNT bundles it is suspected that rapid energy transfer processes take place from semiconducting to metallic SWNTs in bundles of tubes. Lauret *et al.* obtained a relaxation time of 1 ps in the lowest valence and conduction band for bundled tubes [58]. Luminescence corresponding to the E_{11} transition in individual SWNTs is associated with a recombination time of $\sim 20\text{-}180$ ps, depending on temperature and tube diameter [59] and thus corresponds to a process 1-2 orders of magnitude slower than the relaxation time in bundled tubes. This lifetime difference explains the absence of PL in nanotube bundles: metallic SWNTs provide additional and efficient non-radiative decay channels and quench the excitation. In small bundles of only semiconducting SWNTs, a similar energy transfer towards the tube with the lowest band gap is expected. Recent PL experiments on ensembles and individual SWNTs show indications of a 'Förster Resonant Energy Transfer' (FRET) [60, 61].

Absorption spectroscopy can be performed with metallic and semiconducting SWNTs for energies up to E_{33}^S. Higher order transitions are obscured by the tail of a broad π-plasmon resonance, peaking at about 4.5-5 eV [62, 63]. Direct absorption measurements on individual tubes have not been performed yet due to the low absorption cross section. Bundles of SWNTs show broad absorption signals in the energy ranges E_{11}^S, E_{22}^S, E_{11}^M etc. which correspond to the E_{ii} transitions of all chiralities present in the sample, broadened by tube-tube interactions. Only after individualizing the tubes in a solvent/surfactant mixture is it possible to identify the fine structure of these signals

2.4 Optical Properties

now corresponding to the transitions of different chiralities.

2.4.4 Excitons in Carbon Nanotubes

For a qualitative agreement of theory and experiment, tight-binding and zone folding calculations yield quick results which can be interpreted intuitively. However, when it comes to a quantitative comparison, deviations arise which can only correctly accounted for by taking Coulomb interaction into account. The picture used so far was that of individual, non-interacting electrons and holes and is now replaced by interacting electrons and excitons. The importance of many body effects in SWNTs was first noticed in the context of the 'ratio problem' [64, 65]: An experimentally obtained plot of E_{22}/E_{11} vs. tube diameter extrapolated for large diameters yields a ratio of 1.7-1.8. According to one-electron band theory, this ratio should approach 2 for $d \to \infty$. The discrepancy is resolved when considering Coulomb interaction between electrons in general and electron-hole pairs created upon light absorption and bound by Coulomb interaction, so-called excitons. The former effect tends to increase the band-gap and the energy of higher order van Hove singularities due to repulsive interactions, the latter introduces additional states below E_{ii} due to strongly attractive electron-hole interactions. Although these effects are of opposite signs, they do not cancel completely and there still remains a notable blueshift of both E_{11} and E_{22} because the effect on the band gap is larger [57, 66]. This net energy change decreases the $d \to \infty$ limit of the E_{22}/E_{11} ratio from 2 to 1.8.

Generally, excitons are classified into two limiting cases: Mott-Wannier-type [67, 68] and Frenkel-type [69]. Frenkel-type excitons have large binding energies (on the order of 1 eV) and average electron-hole distances of about the lattice constant. They are common in insulators and ionic solids. Mott-Wannier excitons are generally found in 3D bulk semiconductors like Si, Ge and GaAs. Due to a low binding energy (E_b) of only 1 - 50 meV, they can only be observed in low-temperature experiments. Accordingly, the average size of a Mott-Wannier exciton is much larger than the lattice constant. The energy levels (E_n) and binding energy ($E_b = E_g - E_1$) of a Mott-Wannier exciton in a typical bulk 3D semiconductor can be calculated using a modified Rydberg formula, employing the reduced effective mass of electron and hole (μ^*) and the dielectric constant

2 Carbon Nanotube Basics

of the material (ϵ_r):

$$E_n = E_g - \frac{13.6\,\text{eV}}{n^2} \cdot \frac{\mu^*}{\mu_H} \cdot \frac{1}{\epsilon_r^2} \quad \text{with} \quad n = 1, 2, 3... \tag{2.31}$$

where μ_H is the reduced mass of the hydrogen atom. In contrast, excitons in semiconducting carbon nanotubes have a large (calculated) binding energy of up to 1 eV which is mainly due to a confinement effect and a resulting enhancement of Coulomb interaction in these quasi-1D systems [70, 71]. In metallic SWNTs, other conduction electrons shield the attractive interaction and binding energies are much lower. Calculations show that exciton size in z-direction is comparable to the lattice constant a for chiral tubes. Therefore, excitons in most semiconducting carbon nanotubes are rather Frenkel-type than Mott-Wannier-type. Nevertheless, hydrogenic-like exciton levels are found in SWNTs, as well.

Figure 2.8(c) and (d) shows singlet excitonic levels in chiral and zig-zag SWNTs, respectively. They are deduced from wave vector group theory and calculations for some exemplary tubes and may not represent a correct relative ordering for all SWNTs [54–57]. For the symmetry assignment of the exciton wave functions, the effective-mass and envelope-function approximations were used (EMA) [72].

$$\psi^{EMA}(\vec{r}_e, \vec{r}_h) = \sum_{v,c} B_{vc} \phi_c(\vec{r}_e) \phi_v^*(\vec{r}_h) F_\nu(z_e - z_h) \tag{2.32}$$

Within this framework, the exciton wave function ψ^{EMA} is given as a linear combination of only those valence and conduction bands (ϕ_v and ϕ_c, respectively) which contribute to a given van Hove singularity. Although Coulomb interaction mixes all states, this is a good approximation for SWNTs with $d < 1.5$ nm with a large energy separation of the van Hove singularities. However, the approximate exciton wave functions ψ^{EMA} still possess the same symmetry as the full wave functions. The coefficients B_{vc} are dictated by symmetry. $F_\nu(z_e - z_h)$ are envelope functions which localize the exciton in the relative coordinate $z_e - z_h$ along the nanotube axis. The envelope function is either even ($\nu = 0, 2, 4...$) or odd ($\nu = 1, 3, 5...$) under $z \to -z$ operations. For every envelope function F_ν, there exist four (three) excitonic states for chiral (zig-zag) nanotubes: $A_1(0), A_2(0), \mathbb{E}_{m'}(k'), \mathbb{E}_{-m'}(-k')$ [$A_{1u}(0), A_{2u}(0), \mathbb{E}_{m'u}(0)$ for ν even and $A_{1g}(0), A_{2g}(0), \mathbb{E}_{m'g}(0)$ for ν odd].[4] With increasing ν, the energy separation between

---

[4] Note that the values of m' and k' will in general be different for each nanotube and also for each E_{ii} transition.

2.4 Optical Properties

the states as well as the separation within the $\nu = const.$ states decreases, until electron and hole are no longer bound to each other for $\nu \to \infty$. This 'series limit' is depicted as bold lines in Fig. 2.8(c) and (d) and complies with free electron and hole moving in the conduction and valence band of the corresponding van Hove singularity. Hence, the series limit formally corresponds to E_{ii} in the single-particle picture. However, most of the oscillator strength is now transferred to the lowest optically active exciton in each series. They are responsible for the observed main absorption and emission signals in SWNTs and are therefore still labeled E_{ii}.

Although the number of states has greatly increased compared to the single-particle picture, only few excitons can actually be observed experimentally. This can be explained using group theory. For chiral (zig-zag) nanotubes, the ground state transforms as the totally symmetric representation A_1 (A_{1g}) and the interaction between linear polarized light parallel to the SWNT axis and the electric dipole moment as A_2 (A_{2u}). In the dipole approximation, an optical transition between initial $|i\rangle$ and final $|f\rangle$ states is allowed if the direct product between the representations of the dipole moment and the electronic states contains the totally symmetric representation:

$$\mathcal{D}^{|f\rangle} \otimes \mathcal{D}^{\mu} \otimes \mathcal{D}^{|i\rangle} \supseteq A_1 \tag{2.33}$$

and for zig-zag tubes:

$$\mathcal{D}^{|f\rangle} \otimes \mathcal{D}^{\mu} \otimes \mathcal{D}^{|i\rangle} \supseteq A_{1g} \tag{2.34}$$

Additionally, as mentioned earlier, only transitions with $K = 0$ are optically allowed due to linear momentum conservation. Therefore, only A_2 excitons for chiral and A_{2u} excitons for zig-zag SWNTs are bright (i. e. optically active), illustrated in red and blue for the first and second parallel optical transition in Fig. 2.8(c) and (d), respectively.

Even though group theory is useful for qualitatively assigning the symmetry and optical properties (bright/dark) of excitons, only a calculation of energies and transition matrix elements can yield their correct energetic order and oscillator strength. Most calculations on an *ab-initio* level focus on thin armchair or zig-zag tubes to reduce computational effort. Spataru *et al.* and others first use density-functional theory (DFT) and the local density approximation (LDA) to calculate the ground state atomic structure and a mean field description of electronic states. Secondly, they apply one-particle Green's functions (so-called GW approximation) to account for the correct exchange and correlation potential and finally the Bethe-Salpeter equation (BSE) which yields exciton states and optical transition strengths [6, 72]. Such calculations confirm the energetic ordering of excitons in Fig. 2.8(c) and (d). In particular, they predict a dark singlet exciton below the lowest bright singlet exciton for all semiconducting nanotubes. For instance, Spataru et al. calculated energy spacings of $\sim 30\,\text{meV}$ for the (10,0) and (11,0) nanotubes, strongly depending on diameter [73]. In addition to dark singlet excitons levels, there are also optically forbidden triplet excitons which are even lower in energy (another 10-20 meV according to Spataru et al.). Only for $A_{1(u)}$ excitons, singlet and triplet states are degenerate because the exchange interaction vanishes by symmetry [57, 73]. In general, however, there is a lot of disagreement about the energy spacings in the literature. For instance, reports about the energy difference of $A_{2(u)}$ and $A_{1(u)}$ vary between 5 and more than 100 meV. Similarly, triplet excitons are calculated to be between 20 and 300 meV below $A_{2(u)}$ [8, 56, 73-77]. For experimental observations of so-called 'deep' excitonic states (states below the lowest optically active exciton) see Sec. 5.3.

These dark excitons below the lowest bright exciton could explain the low PL quantum yield η observed for semiconducting SWNTs. Ensemble measurements in aqueous dispersions estimate it to be on the order of $\eta = 0.01 - 0.1\%$ [65, 78-80] but Lefebvre et al. reported a quantum efficiency of up to 7% for an individual, suspended semiconducting SWNT [81]. The determination of absolute nanotube quantum yields requires a lot of assumptions like absorption cross sections and the amount of absorbers in the sample (for ensemble measurements). This uncertainty partly accounts for the big spread in η. On the other hand, all carbon atoms of a SWNT are surface atoms and the surroundings have a big influence on the PL. In this respect, individual, suspended tubes without an interacting matrix have less non-radiative decay channels than surface-bound or micelle-encapsulated ones and should exhibit a higher η.

A Kataura plot in the single-particle picture is a diagram of measured or calculated band transitions E_{ii} vs. the diameter of SWNTs. In principle, this picture can be retained for excitons. Although the excitonic picture is more complex, most of the excitonic states are dark or have a very weak oscillator strength. A qualitative reason why most oscillator strength is transferred to the lowest A_2 (A_{2u}) exciton in each band is the symmetry of the envelope function $F_\nu(z_e - z_h)$. For $\nu = 0$ it resembles a Gaussian function and has the same symmetry as a 1D harmonic oscillator for higher ν ($\propto z^\nu \cdot e^{-z^2}$). If an exciton is to decay radiatively, both electron and hole must be found at the same position. For

2.4 Optical Properties

F_ν odd, this probability is very low since the function has a node at $z_e - z_h = 0$ and for $\nu = 2, 4, 6...$ the intensity around $z_e - z_h = 0$ is much lower than for $\nu = 0$ [57]. The Kataura plot can therefore be interpreted as the energy of the bright exciton state with $\nu = 0$ for each band-band transition as a function of tube diameter.

Apart from the ratio problem discussed above, direct experimental indication of excitons in SWNTs was given by two-photon PLE experiments [7, 8]. Absorption and emission in SWNTs is normally a single-photon process but at high excitation power densities, two-photon absorption (TPA) can occur at an appreciable rate. TPA obeys different selection rules than one-photon absorption (OPA). The irreducible representations of the dipole moment, $\mathcal{D}^\mu$ in Eqs. 2.33 and 2.34 are now $A_2 \otimes A_2 = A_1$ for chiral and $A_{2u} \otimes A_{2u} = A_{1g}$ for zig-zag tubes. Thus, TPA will access A_1 and A_{1g} excitons. Together with OPA, exciton binding energies can be estimated but the exact number is model dependent. Wang et al. determined a binding energy of $\sim 400\,\text{meV}$ for SWNTs with 0.8 nm diameter. Similarly, $E_b \approx 300 - 400\,\text{meV}$ was determined for tubes between 0.68 and 0.9 nm in TPA measurements by Maultzsch et al.

Another strong indication that the optical properties of carbon nanotubes are excitonic comes from the experimental observation of exciton-phonon complexes [82–84], demonstrating both the existence of excitons and the central role played by phonons in describing the excitation and recombination mechanisms in SWNTs. In a typical PLE map, weak signals are observed above and below the main E^S_{22} absorption resonances which are attributed to phonon sidebands (as provable by the energy shift of ^{13}C enriched SWNTs compared to pure ^{12}C nanotubes [52]). Qualitatively, the energy separations of these phonon sidebands from the true excitonic transitions (also called zero-phonon lines) are comparable to the energies of prominent Raman active modes. However, the PL sidebands are quantitatively somewhat higher in energy which, according to Ref. [85] is due to the existence of dark excitons roughly 30 meV above the optically active exciton that predominantly couple to the specific phonons. This shift and also the lineshape rule out the possibility of a simple Raman process. Exciton-phonon complexes are quasi-particles, in which electron, hole and phonon form an entity with a specific binding energy, just like electrons and holes in excitons. In contrast, Perebeinos et al. predict no phonon sidebands for pure band-to-band transitions. The Raman modes which can be responsible for the creation of exciton-phonon complexes will be discussed in the next subsection.

2.4.5 Raman Spectroscopy

As mentioned in Sec. 2.4.1, Raman spectroscopy was among the first methods used to study carbon nanotubes, because it requires little or no sample preparation and it can be applied to metallic and semiconducting SWNTs. This subsection covers the most important Raman active modes and what they tell us about the structure of a SWNT. In first-order (= one-phonon) Raman spectroscopy, an incoming photon with (angular) frequency ω_1 and wave vector $\vec{k}_1$ interacts with an electronic excitation and scatters inelastically under emission of a phonon of frequency ω_{ph} ($\vec{q}$), resulting in a photon ω_2 ($\vec{k}_2$). Energy and momentum are conserved:

$$\begin{aligned} \hbar\omega_1 &= \hbar\omega_2 \pm \hbar\omega_{ph} \\ \vec{k}_1 &= \vec{k}_2 \pm \vec{q} \end{aligned} \qquad (2.35)$$

where ± refers to Stokes and anti-Stokes scattering. Optical photons have negligible momenta, thus $\vec{q} \approx 0$. In first-order Raman spectroscopy, only transitions around the Γ point of the phonon dispersion relation of SWNTs are observed. The phonon dispersion relation $E(\vec{q})$ of carbon nanotubes can be constructed similarly to the electronic dispersion relation [$E(\vec{k})$] by applying the zone-folding approximation to the 2D phonon dispersion relation of a graphene layer [21]. If, apart from the ground state, only virtual excited states are involved in a first-order Raman scattering process, it is termed non-resonant [see Fig. 2.9(a)]. If the incoming (ω_1) or the outgoing photon (ω_2) hits a real electronic (excitonic for SWNTs) state of the system, single- (incoming or outgoing) resonant scattering occurs [Fig. 2.9(b) and (c)]. The scattering amplitude and therefore the Raman intensity is strongly enhanced in a resonant process, such that even Raman spectra of individual SWNTs can be observed. Raman spectroscopy of nanotubes is always dominated by signals of tubes in resonance with the incident laser light. Moreover, since a resonant process always includes the creation of an exciton, the same selection rules as for absorption apply and PL signals are often observed together with Raman bands and vice versa. Resonance Raman Spectroscopy (RRS) can be used to determine chiralities present in an ensemble if the incident laser light is tuned. This procedure is comparable to PLE spectroscopy although the small Stokes or anti-Stokes shift in Raman spectroscopy creates technical difficulties.

Double-resonant processes as depicted in Fig. 2.9(d) and (e) should be even more enhanced. However, (d) includes elastic scattering by a defect (dashed arrow) and (e)

2.4 Optical Properties

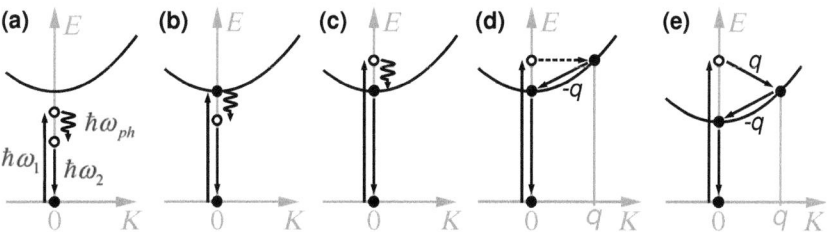

Figure 2.9: Schematic diagrams of Stokes Raman scattering processes in SWNTs. Solid and open circles denote real and virtual states, respectively. (a)-(d) represent first-order (= one-phonon) scattering and (e) is a second-order (= two-phonon) process. (a) Non-resonant, (b) incoming single-resonant, (c) outgoing single-resonant, (d) and (e) double-resonant. The dashed arrow in (d) indicates elastic scattering by a defect in the crystal. Anti-Stokes scattering can be deduced by considering the time-reversed processes. Modified from Ref. [86].

involves consecutive scattering with two phonons which is not very likely in perfect, infinitely long tubes. Whether they can be observed in SWNTs thus depends on the product of enhancement factor and probability. It turns out that both, defect-induced double-resonant process and second-order scattering, (also called 'overtones') are found in SWNTs. In the following, the most important Raman active modes in SWNTs will be discussed.

Radial Breathing Mode The radial breathing mode (RBM) is a low frequency, first-order, totally symmetric Raman mode between 100 and 400 cm^{-1}, depending on the tube diameter. All carbon atoms move in phase in the radial direction creating a breath-like vibration of the entire tube. Obviously, this mode has no counterpart in graphene, but it is well known in Raman spectroscopy of fullerenes. Theoretical and experimental results show that the RBM frequency (ω_{RBM}) is inversely proportional to the diameter d of the SWNT:

$$\omega_{RBM} = \frac{A}{d} + B \quad (2.36)$$

The constants A and B are being determined experimentally and vary with environment:

- Suspended tubes: $A = 204\,\text{cm}^{-1}\text{nm}$, $B = 27\,\text{cm}^{-1}$ [87]
- SWNTs on a SiO$_2$ substrate: $A = 248\,\text{cm}^{-1}\text{nm}$, $B = 0$ [88]
- Bundled nanotubes: $A = 239\,\text{cm}^{-1}\text{nm}$, $B = 0$ [89]

- SWNTs dispersed in aqueous solution: $A = 218\,\mathrm{cm^{-1}nm}$, $B = 16\,\mathrm{cm^{-1}}$ [90]

The diameter dependent frequency of the RBM is the reason why the spectra of this mode are largely used for the characterization of the diameter distribution in a carbon nanotube sample. In so-called RBM resonance profiles, the RBM frequency shift as a function of excitation energy is measured and the evaluated E_{ii} and ω_{RBM} values can be used to assign chiral indices to the signals [90, 91].

D-Band The D-mode is a one-phonon, second-order, double resonance Raman process, already depicted in Fig. 2.9(d). It includes the elastic scattering of an exciton by a defect and is common to all sp^2-hybridized disordered carbon materials (e.g. graphene or amorphous carbon). It is observed at around $\omega_D = 1350\,\mathrm{cm^{-1}}$ and becomes active in carbon nanotubes due to the presence of defects, such as impurities (e.g. dopants), vacancies, 7-5 pairs (a sidewall defect comprising of a carbon heptagon adjacent to a pentagon), molecules linked to the SWNT sidewall, sp^3-carbon or nanotube ends. The intensity of the D-mode scales with the defect concentration [86, 92] and is thus widely used for the characterization of the quality of SWNT samples. The exact position of the band for a given nanotube depends on the excitation laser energy. Upon an increase of excitation energy, ω_D has to increase as well, in order to fulfill the double resonance condition. For a given laser energy, different chiralities of isolated SWNTs exhibit different ω_D for the same reason. The observed band is therefore a sum of different features originating from different chiralities [86, 93].

G-Band The G-modes are the highest energy, first-order modes in SWNTs and are also known as high-energy or tangential modes (HEM or TM, respectively), because they are identified with a tangential vibration of adjacent carbon atoms, parallel (so-called longitudinal optical phonon, LO) and perpendicular (transverse optical phonon, TO) to the tube z-axis. The G-band in SWNTs is named after the G-band in graphite where it exhibits a single Lorentzian feature at $\sim 1582\,\mathrm{cm^{-1}}$.[5] Due to the quantum confinement of the phonon wave vector along the SWNT circumferential direction and zone folding, theory predicts six G-modes for chiral, and three for achiral (zig-zag and armchair) nanotubes, with chirality and polarization dependent intensities [94, 95]. They group into two separately observable components, G^- and G^+, at $\sim 1570\,\mathrm{cm^{-1}}$ and $\sim 1590\,\mathrm{cm^{-1}}$, respectively. The G^- feature is slightly diameter dependent and also sensitive to whether the tube is semiconducting or metallic, showing an asymmetric Breit-Wigner-Fano (BWF)

---

[5]This is also the only G-band feature in MWNTs.

lineshape and a lower frequency (at constant diameter) for the latter [96]. The G^+ feature is independent of diameter.

G'-Band The G'- or D^*-band is the most intense two-phonon, second-order, double resonance Raman feature present in SWNTs. It is the overtone of the D-band, thus $\omega_{D^*} = 2\omega_D \approx 2700\,\text{cm}^{-1}$. Instead of involving a phonon and a defect as for the D-band, another phonon is responsible for momentum conservation in the double resonance process, see Fig. 2.9(e). The G'-band in graphene is a single Lorentzian feature and was shown to depend on the number of graphene layers in a graphite sample with a few layers (< 5 layers) [97, 98]. In SWNTs, there are sometimes two close features observed for the G'-band which indicates the resonance with two different van Hove singularities [99].

2.4.6 Conclusion to Spectroscopic Methods

The spectroscopic methods presented in the last subsections are very useful to characterize a carbon nanotube sample in a quick and nondestructive way (as opposed to e.g. diameter statistics obtained by high-resolution transmission electron microscopy, HRTEM). Most importantly, they provide the possibility to assign (n_1, n_2) chiral indices to spectroscopic signals. PLE spectroscopy was the first method which enabled Bachilo et al. in 2002 to successfully assign (n_1, n_2) to experimentally measured E_{11} and E_{22} transition energies by comparison with a theoretical Kataura plot [2]. Theoretical and experimental values of E_{ii} showed a notable disagreement as they used tight-binding calculations which did not include excitonic effects. However, the expected pattern was similar and by using the RBM of some tubes, they were able to 'anchor' the assignment. Subsequently, they extrapolated their empirical findings to a wide range of semiconducting SWNTs [100].

As mentioned earlier, RBM resonance profiles can also be used for a chirality assignment, especially for the assignment of metallic SWNTs [88, 90, 101, 102]. However, this is technically more challenging than PLE spectroscopy because of the small Raman shift and the need for a lot of tunable lasers.

Closely related to the question of assignment is the issue of abundances of individual chiralities in SWNT samples. Intensities in PLE contour maps are convoluted by the absorption cross section σ and quantum efficiency η which are believed to depend on chirality. Thus, for a determination of abundances via PL spectroscopy, η and σ have to be known. The exact relation of η and σ to the structure of SWNTs is still part of ongoing research [103–105].

In this work, we show that the counting of individual SWNTs by µ-PL(E) spectroscopy is also a possible route to determine chiral abundances of ensembles (see Sec. 5.5). µ-PL(E) spectroscopy was also used to study environmental and excitonic effects (Sec. 5.1, 5.2 and 5.3), the homogeneity of chiral indices of ultralong nanotubes and the influence of strain applied by atomic force microscopy (AFM) (Sec. 5.4). In addition, µ-Raman spectroscopy helped in the optimization of the chemical vapor deposition (CVD) process and was used to check the quality and diameter distribution of samples. The next chapter deals with the details of this process and synthesis methods in general.

2.5 Synthesis Methods of Carbon Nanotubes

The synthesis of carbon nanotubes currently faces four main challenges [6]: 1. Mass production, i.e. the development of low-cost, large-scale processes for the synthesis of high-quality nanotubes, including SWNTs; 2. Selectivity, i.e. the control over chirality, diameter and length; 3. Organization, i.e. control over placement and orientation on a flat substrate; and 4. Mechanism, i.e. an understanding of the processes of nanotube growth.

There has been progress in each of these fields and covering all of them would certainly go beyond the scope of this thesis. Therefore, this chapter gives a quick review of different approaches for the synthesis of carbon nanotubes. The techniques can be divided into three main classes:

- Arc discharge
- Laser ablation/vaporization
- Chemical vapor deposition (CVD)

All procedures use metal catalysts, especially transition metals like iron, cobalt, molybdenum, nickel and yttrium or mixtures thereof for the synthesis of SWNTs. Historically, the arc discharge method was the first to successfully produce bulk amounts of SWNTs. Together with the laser vaporization technique, it belongs to the high temperature synthesis techniques with $T \gtrsim 3500\,°C$. At such elevated temperatures, the precursor material (graphite and metal catalysts) is vaporized and SWNTs start to nucleate on metal particles upon cooling. CVD is a low or medium-temperature synthesis technique with $T \approx 600-1200\,°C$. The lower temperature is favorable regarding industrial implementation and mass production, direct growth of SWNTs on integrated circuits and also regarding selectivity. Thus, CVD methods have gained much attention in the last few years and a substantial part of this thesis was attributed to the design, construction and optimization of our own CVD set-up (see Sec. 3.3) Tubes from this set-up are mainly used throughout this work.

2.5.1 Arc Discharge

In an arc discharge reactor, two electrodes made of graphite are brought closely together in an inert atmosphere (typically helium or argon) at reduced pressure (typically

∼500 mbar). Without a catalyst, Fullerenes and MWNTs are formed, when a DC discharge (of about 300 A per cm^2 of electrode area) is ignited between the electrodes. In order to make SWNTs, the center of the anode is usually filled with a mixture of catalyst and graphite powder. During the process, SWNTs are formed on the cathode and the anode is eroded (consumed). Yields estimated from scanning electron microscopy (SEM) images between 10-20 wt% of SWNTs (referred to the total amount of evaporated carbon) with average diameters of ∼1.4 nm are observed [106].

2.5.2 Pulsed Laser Vaporization

In a pulsed laser vaporization (PLV) set-up, focused laser pulses evaporate a target made of sintered graphite and catalyst powder. The target is located inside a furnace at 1150 - 1200 °C. A constant flow of argon at ∼500 mbar cools the generated plume and carries the nanotubes out of the furnace where they can be collected [107]. Thess *et al.* reported an estimated SWNT yield of more than 70 wt% (relative to the total evaporated carbon) and a medium diameter of 13.8±0.2 Å with this method [108].

The PLV set-up at the University of Karlsruhe utilizes a pulsed Nd:YAG laser (Continuum 'Powerlite', Q-switched, 1064 nm, 0.5 J/pulse, 30 Hz) which is focused on a purchased rod-like target with a length of 50 mm and a diameter of 10 mm. The rod contains 1.2 atom% cobalt and nickel, respectively and is located in a tube furnace. The laser beam hits the target at a 90° angle through a hole in the tube furnace, corresponding to a T-like configuration [109]. The as-produced material typically contains up to 50 wt% SWNTs with a diameter between 0.9 nm and 1.4 nm as deduced from PLE measurements [30]. PLV tubes from this set-up were used for projects which have already been discussed in Ref. [110] and [111], and are therefore not part of this thesis.

2.5.3 Chemical Vapor Deposition

In contrast to the arc discharge or PLV process, the carbon feedstock in CVD is not a carbon plasma but an 'activated' carbonaceous gas like methane, ethylene, acetylene or carbon monoxide. Also hydrocarbons or alcohols which are liquids under ambient conditions are used. Very often, auxiliary gases like hydrogen, argon, nitrogen or ammonia are added. The activation of the molecules is achieved using a variety of methods, which can be roughly summarized as (i) thermal and (ii) plasma enhanced (PE) CVD.

2.5 Synthesis Methods of Carbon Nanotubes

The former applies higher temperatures to excite the precursors, whereas in PECVD electromagnetic (DC or AC) fields are used to ignite and sustain electrical discharges. CVD tubes in this thesis were produced by thermal CVD only.

Thermal CVD can further be divided into methods employing a floating or a supported catalyst. For example, the HiPco process developed by Smalley *et al.* is a floating catalyst CVD method [112]. HiPco stands for **high-p**ressure carbon monoxide (**CO**) and refers to a process, in which SWNTs grow in a flow of carbon monoxide at high pressure (30 - 50 bar) and temperatures of 900 - 1100 °C on catalytically active iron clusters in the gas phase. The clusters are formed *in situ* from iron pentacarbonyl, $Fe(CO)_5$, added to the CO flow. Due to the high temperatures, $Fe(CO)_5$ decomposes and forms iron clusters. The clusters catalyze the formation of SWNTs through the disproportionation of CO on their surface (Boudouard equilibrium). SWNTs are cooled and removed from the hot reaction zone by a continuous gas flow. The raw material contains SWNTs in a purity of up to 50 wt% and diameters between 0.7 and 1.1 nm [113].

The CVD reactor built as part of this thesis uses only supported catalysts. The synthesis is therefore essentially a two-step process consisting of a preliminary catalyst preparation step followed by the actual synthesis of the nanotubes. The advantage of this approach is a high control over the placement and orientation of the tubes: by patterning the catalyst, carbon nanotubes can self-assemble into predefined, aligned structures during growth. Highly-oriented growth morphologies are either **v**ertically **a**ligned carbon **n**anotube **a**rrays (VANTAs, also called nanotube 'forests') or horizontally aligned arrays. The former is only possible if the density and the growth rate is high enough so that adjacent tubes can interact via van der Waals forces to gain rigidity, which allows them to self-orient and grow perpendicular to the substrate. The latter needs an aligning force such as an electric or magnetic field, the gas flow or the surface. Otherwise, a random orientation is obtained. Low-density, horizontally aligned parallel SWNTs are ideally suited for individual nanotube μ-PL(E) or μ-Raman spectroscopy, because the polarization of the excitation light only needs to be adjusted once (i. e. mainly parallel to the tube). The growth of VANTAs can also be favorable for spectroscopy on the single tube level, if they mainly consist of MWNTs with only a low density of SWNTs suspended in between (see Sec. 5.1).

It is generally accepted (and experimentally shown, e. g. by controlled atmosphere electron microscopy, i. e. CVD inside a HRTEM) that nanotubes in medium temperature

2 Carbon Nanotube Basics

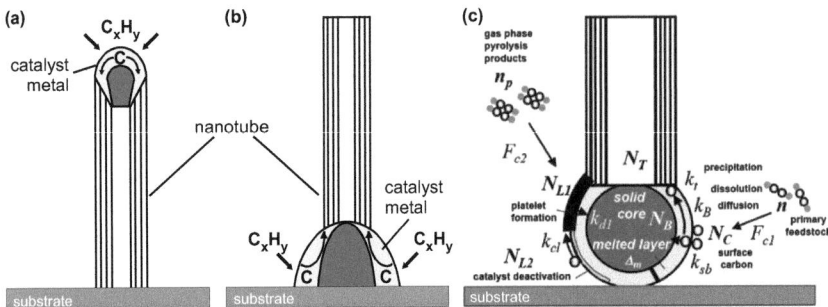

Figure 2.10: Carbon nanotube growth mechanisms. **(a)** Tip growth due to weak catalyst-support interaction. **(b)** Base growth due to strong catalyst support-interaction. **(c)** Detailed kinetic model for acetylene as carbon feedstock. **(Right)** (1) impingement of C_2H_2 molecules onto the catalyst particle surface; (2) chemisorption and catalytic decomposition on the surface of the catalyst particle; (3) surface-bulk penetration of carbon atoms with the rate constant k_{sb}; (4) formation of a disordered surface layer with thickness Δ_m; (5) diffusion of carbon atoms channeled by the disordered layer; (6) precipitation of carbon species into a nanotube with rate constant k_t. **(Left)** Mechanisms that finally stop the decomposition of C_2H_2 and terminate the growth like **(top)** growth of a carbonaceous layer due to gas-phase pyrolysis products or **(bottom)** catalyst deactivation, e. g. due to oxygen reduction at higher temperatures and formation of an inactive layer. From Ref. [114].

CVD grow by the extrusion of carbon, dissolved in a metallic catalyst particle that is oversaturated in carbon, and that the catalyst particles promote tip growth or base growth (or both) depending on the interaction strength between the catalyst particles and the substrate as depicted in Fig. 2.10(a) and (b) [115]. This model is similar to the VLS model (**v**apor-**l**iquid-**s**olid), developed in 1964 to account for the growth of silicon whiskers [116]. A consequence of this model is the relation between the size of the catalyst particle and the diameter of the nanotube. It was shown in many experiments, and also in this thesis (see Sec. 4.1), that the diameter of a catalyst nanoparticle is close to the diameter of the tube [117–119].

However, the exact mechanisms depend crucially on parameters like catalyst and catalyst pretreatment, temperature, carbon feedstock, mass flow, gas composition etc. Fig. 2.10(c) shows a kinetic model deduced by Puretzky *et al.* from *in situ* measurements of VANTA growth [114]. They used a triple layer of evaporated aluminum, iron and molybdenum (with thicknesses of 10, 1 and 0.2 nm, respectively) on a silicon substrate as catalyst and an argon/acetylene/hydrogen gas mixture.[6] The process that leads to the formation of carbon nanotubes is shown on the right side of Fig. 2.10(c). C_2H_2 feedstock molecules at concentration n collide with the surface of a catalyst nanoparticle, providing a flux of carbon-containing molecules, F_{c1}, to the surface of the nanoparticle. A small fraction of these molecules bonds to the nanoparticle surface due to chemisorption and catalytically decomposes, producing the starting source of carbon atoms, N_C, for nanotube growth. Carbon atoms are dissolved with the rate constant k_{sb}, forming a highly disordered 'molten' layer of thickness Δ_m on the surface of the catalyst nanoparticle. Δ_m determines the number of walls of the carbon nanotube which is precipitated with a rate constant k_t from N_B dissolved carbon atoms. Competing catalyst-poisoning mechanisms [left side of Fig. 2.10(c)] are e. g. the growth of a carbonaceous layer containing N_{L1} carbon atoms, with the overall rate constant k_{cl}. It is assumed that acetylene forms gas-phase pyrolysis products of concentration n_p and that the flux of these products, F_{c2}, leads to the formation of a carbonaceous layer which in turn can be partly dissolved by the catalyst nanoparticle with the dissolution rate constant k_{d1}. Other deactivating processes like formation of iron carbide, FeC_3, change in the oxidation state or diffusion of Si atoms from the substrate into the nanoparticle are summarized via the introduction of an inactive catalyst layer, N_{L2}. By solving the rate equations associated with the above-mentioned processes, Puretzky *et al.* were able to fit *in situ* measured growth rates, temperature dependences, terminal lengths and acetylene flow rate dependences for VANTAs. The next paragraph gives a short historical abstract of VANTA growth.

Vertically aligned arrays of carbon nanotubes The first VANTAs employing a mesoporous silica template (which also served as a substrate), iron as a catalyst and acetylene as feedstock gas were grown in 1996 by Li *et al.* [120]. The generated multi-walled tubes grew up to 100 µm at a rate of ∼ 25 µm/h, covering a maximum area of several square millimeters. Two years later, Ren *et al.* grew arrays of MWNTs up to 20 µm in height by PECVD without the use of a template. Instead, a 40 nm-thick layer of radio frequency (RF) magnetron sputtered nickel on glass served as a catalyst. They observed growth rates of 120 µm/h and the tubes covered an area of several square

---

[6]These parameters are close to the ones we used in Sec. 4.1.

centimeters [121]. Fan and coworkers patterned the iron catalyst via electron beam evaporation through shadow masks which caused the nanotubes to vertically align along the pattern [122]. A big breakthrough was the introduction of so-called 'supergrowth' by Hata *et al.* [123]. Not only were they able to grow 2.5 mm high 'forests' within only ten minutes, corresponding to an increase of average growth rate of more than two orders of magnitude compared to former experiments, the VANTAs also consisted of SWNTs (and DWNTs) without carbonaceous overcoating with an average diameter of 3 nm. Most growth methods are limited by the low activity and the short lifetime of the catalyst. Hata and coworkers solved this problem by introducing a small (in the ppm range) amount of water vapor which presumably etches away or prevents amorphous carbon coverage of catalyst nanoparticles and of the SWNTs [124]. Large-diameter SWNTs made by the 'supergrowth' process are not suited for spectroscopic PL examination because of IR absorption in most media capable of SWNT dispersion and because of relatively insensitive detectors in this wavelength range. Maruyama *et al.* used ethanol as carbon feedstock to grow VANTAs with SWNTs of 1 - 2 nm diameter [125]. Although these SWNTs did not reach the same growth rate and had terminal lengths of only several micrometers, they were ideally suited for cross-polarized RRS experiments [50].

The growth of VANTAs occurs predominantly via a base growth mechanism. This can be shown because most VANTAs are easily removed from the substrate e.g. with a razor blade and the catalyst layer remaining on the substrate can be reused several times [123]. Other evidence like the distribution of catalyst material along the z-axis or the direct laser ablation of the VANTAs tips during growth also show that the nanotubes grow mainly from their bases [126]. This is especially the case for evaporated and sputtered catalyst layers. However, for the growth of very long, (several centimeters) horizontally aligned nanotubes, a tip growth mechanism with a partially floating nanotube is assumed. A summary of the development of this growth morphology is presented in the next paragraph.

Horizontally aligned carbon nanotubes In 1998, Kong and coworkers were the first to grow horizontal SWNTs on silicon substrates patterned by electron beam (e-beam) lithography with micrometer scale islands of catalytic material [127]. The nanotubes were not aligned and had lengths up to 20 µm. By applying electric fields, the same group succeeded in aligning SWNTs across lithographic ridges on a dielectric substrate in 2001 [128]. In the absence of an electric field, the SWNTs grew into ran-

2.5 Synthesis Methods of Carbon Nanotubes

domly suspended networks. Similar field-directed growth was observed by Joselevich *et al.* who also successfully used 'nanotube epitaxy' i.e. the growth of SWNTs along well-defined crystal surfaces for nanotube alignment [129, 130]. Another promising approach to orient SWNTs is the use of the feeding gas to align SWNTs along the flow direction. This was first discovered by Liu and coworkers [131–133]. By employing a fast-heating CVD method, in which the Si wafer patterned with catalyst is heated quickly to the reaction temperature by inserting the substrate into the center of a furnace, most resulting SWNTs are long (up to a centimeter scale) and well-oriented after a growth time of ~ 10 minutes. The fast-heating is believed to cause convection of the gas flow due to the temperature gradient between substrate and feeding gas. This convection lifts the nanotubes upward and keeps them floating until they are caught by the aligning laminar flow. Once they grow longer, they probably stick to the surface due to van der Waals forces. If they were in fact growing from the base, touching the surface would likely terminate their growth or produce a lot of loops and wiggles close to the catalyst pad. However, this is not observed. Thus, the picture of a tip-growing SWNT, partly bound to the surface and partly floating ('kite-mechanism' [134]) is most likely for such very long horizontally aligned SWNTs. More recently, methods without the need of the fast-heating step were developed [15]. Kim and coworkers for example demonstrated the use of a small quartz tube inside a large tube to achieve better alignment [135]. Li *et al.* reached alignment via a very low feeding gas flow [136]. Horizontally aligned arrays of SWNTs from the group of Li were also used in Sec. 5.4.4 in this thesis.

CVD in general is a versatile method with which both the growth of bulk amounts, and individual, well-organized structures of SWNTs can be realized. The vertically and horizontally aligned arrays presented in this subsection are typical examples. However, the involved growth mechanisms, especially the nucleation of SWNTs out of a catalyst particle are still subject of intense theoretical and experimental research. A detailed understanding of these mechanisms is probably the key towards chirality selective synthesis of carbon nanotubes in the future.

3 Experimental

In the following three sections, detailed descriptions of the photoluminescence microscope, the Raman microscope and the chemical vapor deposition reactor are given. Apart from standard scanning electron microscopy (SEM) and atomic force microscopy (AFM), these were the main instruments used throughout the present work.

3.1 Confocal Photoluminescence Microscope

3.1.1 Overview

Spatially resolved photoluminescence (PL) spectroscopy requires coupling of a spectrometer with a microscope. For the investigation of individual SWNTs or small ensembles thereof, the standard body of an upright, technical, infinity-corrected microscope (Zeiss, 'Axiotech vario') was modified to a confocal near infrared (NIR) photoluminescence microscope. Reasons for choosing this type of microscope are two-fold. Firstly, due to its stand height of 380 mm, even bulky and heavy devices like a cryostat or a Diamond Anvil Cell on a (x, y)-translation stage fit underneath the objective. Secondly, the highly modular design allows the construction of customized add-ons, enabling confocal and NIR PL operation. Figure 3.1 shows a sketch of the whole set-up composed of the excitation light sources (tunable lasers), microscope and spectrometer and detector. Light is coupled into and out of the microscope via optical fibers. In addition to the PL measurement mode, the microscope can be used as a normal light microscope, too. In the following, a detailed description of all components will be given. Numbers in brackets refer to Fig. 3.1.

3 Experimental

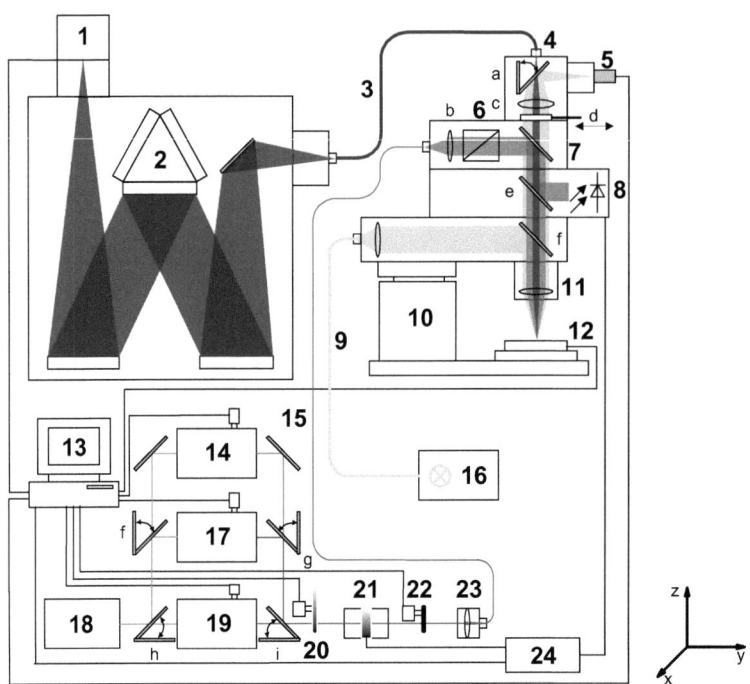

1	InGaAs photodiode array	13	computer
2	spectrograph/ triple grating turret	14	dye laser
3	multi-mode glass fiber	15	single-mode photonic fiber
4	adjustable confocal pinhole	16	cold light source
5	b/w video camera	17	Ti:sapphire laser II
6	polarizing cube	18	argon ion laser
7	dichroic beamsplitter	19	Ti:sapphire laser I
8	Si-photodiode	20	filter wheel
9	optical fiber	21	laser power stabilizer
10	manual focusing stage	22	laser shutter
11	objective	23	fiber coupler
12	piezo scanning stage	24	laser power controller
a-i	flip mirrors, lenses and filters described in the text		

Figure 3.1: Sketch of the photoluminescence microscope. Yellow color represents broadband visible light, laser excitation light between 600 and 1000 nm and PL emission light between 1000 and 1600 nm is drawn in red and dark-red, respectively.

3.1.2 The Laser System

Carbon nanotubes optimally require a broad band of excitation energies, ranging from visible to NIR, as opposed to standard fluorescence spectroscopy where only one excitation wavelength is typically used to characterize a sample. Therefore, three different tunable continuous wave (cw) lasers [Spectra Physics; (14), (17) and (19) in Fig. 3.1], covering together approx. 600 - 1000 nm are used in this set-up. (14) is a dye laser (model 375B), using a solution of DCM in ethylene glycol as active laser medium. (17) and (19) are Ti:sapphire lasers (model 3900S) equipped with different mirror sets. (18) is a 10 W argon ion laser (Spectra Physics, 'Beam Lok 2060') serving as pump source for the above-mentioned lasers, guided by the flip mirrors (h) and (f). All tunable lasers were originally equipped with a manual micrometer screw for wavelength control. It rotates a birefringent filter inside the laser cavity, enabling different frequencies to overcome the lasing threshold. These screws were replaced by motorized actuators (Physik Instrumente) and connected to a computer (13) to provide automated measurements. Flip mirrors (f)-(i) are also equipped with electric servos to quickly switch between different lasers. These are not computer-controlled because switching lasers requires a warm-up period and sometimes additional fine tuning of the beam path.

With 7 W pump power from the argon ion laser operated in the all-lines mode (mainly 488 and 514 nm), the dye laser has a tunable wavelength range of 613 - 707 nm; Ti:sapphire lasers I and II encompass 695 - 865 nm and 833 - 988 nm with 9 W pump power, respectively. The effective spectral range of the laser system is therefore 613 - 988 nm. Output powers vary between 1.8 W and 20 mW ($\sim$ 100:1) for different wavelengths. It is highly desirable to excite nanotubes with a constant power irrespective of the wavelength. This is done in real-time by a laser beam power stabilizer (21) and -controller (24)(BEOC). The laser power controller (LPC) monitors the current of a Si-photodiode (8) which is irradiated by light taken from the main beam via a thin fused silica beam splitter (e). The reflectivity of fused silica is approximately wavelength-independent (between 600 and 1000 nm) whereas the current of the photodiode is wavelength sensitive. The LPC is connected to the computer (13) via RS-232 and receives the current wavelength and the desired output power. It then corrects the current of the photodiode according to the wavelength, calculates the desired attenuation and adjusts the laser power stabilizer (21) accordingly. The laser power stabilizer (LPS) is a liquid crystal modulator directly positioned in the beam path. It has an optimal dynamic range of

3 Experimental

~40:1. In order to reach a 100:1 ratio, an additional filter wheel (20) on a computer-controlled rotary stage (Physik Instrumente) is installed in front of the LPS. The LPC compensates for power fluctuations resulting from a change in the wavelength and also for temporal fluctuations.

Laser light is coupled into a single-mode photonic crystal fiber [Crystal Fibre, (15)] via a fiber coupler [Schäfter+Kirchhoff, (23)]. Standard glass single-mode fibers are only suited for single laser lines or a limited wavelength range. In addition, they show fluctuations in transmission and polarization angle upon bending or twisting, impeding reproducible results. The photonic crystal fiber (PCF) is specified to preserve a TEM_{00} single-mode transmission and linear polarization over our entire wavelength range. Its numerical aperture (NA) increases from 0.11 to 0.18 in the same range, requiring some adjustments on the fiber coupler (23) and the collimating lens (b) for different lasers. As already mentioned in chapter 2.4, optical properties of SWNTs depend crucially on the polarization of incident light relative to the nanotube axis. Therefore, linearly polarized light irrespective of wavelength is preferred. Light exiting the PCF shows TEM_{00} single-mode behavior as confirmed by the eye over the entire visible range (600 - 750 nm) and a very good bending stability. However, a rotation of the linear polarization vector of up to 30° and ratios of $I_\parallel$ to $I_\perp$ from 7:1 to 50:1 for different wavelengths led to the incorporation of a polarizing cube (6). This reduces the fluctuations to 2° and constrains $I_\parallel/I_\perp$ after the objective to ~50:1, independent of the wavelength. The reason why the PCF does not maintain polarization as specified probably results from the fact that the fiber coupler (23) does not allow precise alignment of polarization-maintaining fibers.

3.1.3 The Photoluminescence Microscope

The PL microscope consists of a manual focusing stage (10) combining 3 levels of focusing: (i) a crank on the top of the stand which moves the microscope body up to 22 cm, coaxial handwheels (ii) for coarse focusing and (iii) for fine focusing (2 mm and 0.2 mm per turn, respectively), both enabling a total travel of 15 mm. Additional focusing without touching the microscope is done via a capacitively-controlled, three-axis (x, y, z) piezo scanning stage, on which a sample is placed [Physik Instrumente, (12)]. It is operated in closed loop and provides a resolution better than 5 nm at a full scanning range of 200 x 200 x 20 µm. The piezo stage is connected to the computer (13) and is itself mounted on a manual (x, y)-translation stage with 25 x 25 mm positioning range

3.1 Confocal Photoluminescence Microscope

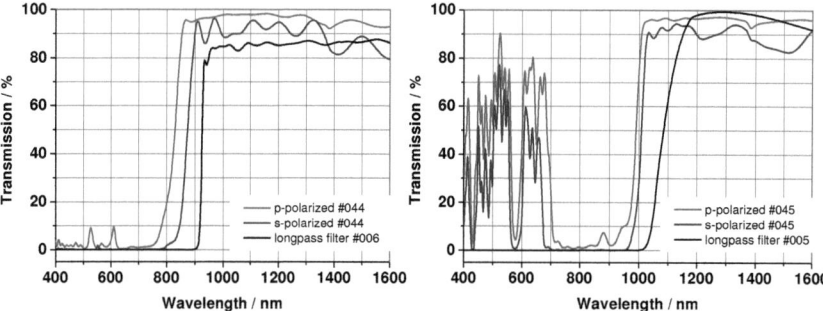

Figure 3.2: Transmission spectra of combinations of beamsplitters and longpass filters. Optical elements were numbered for clear reference. **Left:** Beamsplitter #044 and longpass filter #006 allowed excitation up to 800 or 860 nm, depending on the polarization. Onset of detectable PL is at ∼ 950 nm. **Right:** Beamsplitter #045 and longpass filter #005 can be used up to 975 or 995 nm excitation. Onset of detectable PL is at ∼ 1120 nm. High transmission between 400 and 700 nm improves image brightness in optical mode.

(Newport). Besides fine focusing and moving, the piezo stage enables raster-PL-scanning of the sample (see Sec. 3.1.6). Operation of the instrument includes two different modes:

PL measurement mode In the PL measurement mode, light leaving the PCF is collimated by an achromatic lens (b) [adjustable in (y)-direction for alignment], polarized in (x) or (z)-direction with (6) and reflected downward in $(-z)$-direction by a dichroic beamsplitter [LOT-Oriel, (7)]. The dichroic beamsplitter is positioned at an angle of 45° and placed on a modified mirror mount for adjustment. Beamsplitter (e) extracts a small fraction of light to the photodiode (8) for power measurement. Half-mirror (f) is moved out of the beam path and the cold light source [Schott, halogen lamp with an integrated IR filter, (16)] is turned off. Both beamsplitter (e), photodiode (8) and mirror (f) are drawn above each other in Fig. 3.1 for the sake of clarity. In fact, they are mounted in a horizontal [traveling in (x)-direction] slider next to each other, so that switching between them simply requires moving the slider. The excitation light is then focused onto the sample by an objective (11) (see Tab. 3.1). NIR-Photoluminescence from the sample is collected by the same objective in a reflective geometry. PL passes beamsplitters (e) and (7) with only minor losses and also longpass filter (d) which blocks residual excitation light.

3 Experimental

Figure 3.2 shows transmission spectra of two suitable combinations of dichroic beamsplitter (7) and longpass filter (d). The combination of beamsplitter #044 and filter #006 is used for dye- and Ti:sapphire laser I and is suited for small diameter SWNTs (e.g. HiPco material), which have higher excitation and emission energies than large diameter SWNTs. For #045 and #005, the band edge is red-shifted and coincides with the range of Ti:sapphire laser I and II. This combination is used for larger diameter tubes (e.g. PLV or CVD-grown SWNTs). Figure 3.2 also shows the influence of s- and p-polarized light on the transmission of beamsplitters #044 and #045, respectively. 's' and 'p' refer to a linear polarization perpendicular ('senkrecht') and parallel to a plane generated by the surface normal of the beamsplitter and the incident beam. On the one hand, p-polarization yields better performance for NIR transmission but on the other, the shift of the band edge reduced the usable bandwidth of the Ti:sapphire lasers. In this work, mostly s-polarization was used but could be changed by rotating the polarizing cube (6) by 90°. With flip mirror (a) moved out of the beam, an achromatic tube lens (c) focuses the PL on the entrance of a multi-mode glass fiber with a diameter of 400 µm. The entrance of the fiber serves as a confocal pinhole and is attached to a customized mount which allows adjustments along x, y and z axes. Adjusting the position of the pinhole is critical for intensity optimization of the signal. A 400 µm-sized pinhole is too large for a distinct improvement of resolution compared to standard microscopy [137]. Smaller diameter fibers (200, 100, 50 and 25 µm) were available but hardly used because of too weak signals. Therefore, the resolution given in Tab. 3.1 is limited by the objective alone and not further improved due to confocality. Nevertheless, the arrangement reduces stray light and thereby increases sensitivity.

Optical mode   In the optical mode, light from the cold source [Schott, (16)] enters the microscope through an optical fiber (9) and is used in the standard Köhler scheme (utilizing an aperture- and field iris diaphragm as well as several lenses, not shown) to illuminate the sample through the objective, with half-mirror (f) at an angle of 45°. In this mode, laser light is blocked by a computer-controlled shutter (22). Back-scattered light is collected by the objective and passes through (with losses) half-mirror (f) and dichroic beamsplitter (7). The quartz beamsplitter (e) and laser filter (d) are moved out of the optical path and flip mirror (a) guides the light into a black and white eyepiece camera (5), connected to the computer (13). In the current set-up, the dichroic beamsplitter (7) cannot be removed from the optical path. It is, therefore, important

3.1 Confocal Photoluminescence Microscope

Table 3.1: List of infinity-corrected objectives used for PL measurements on SWNTs. Resolution corresponds to fwhm of the theoretical Point Spread Function (PSF)[138].

Company	Model	Mag./N.A.	Working distance (mm)	Lateral res. (µm)	Remarks
Olympus	MPlan 100xIR	100x/0.95	0.3	0.4	for NIR observation
Nachet	PL.FL LD	40x/0.6	3	0.7	multi-diel. coating 800 - 1600 nm
Ealing	IR Achromatic	15x/0.4	10	1	corrected for 0.85 and 1.3 µm
Leica	HCX PL APO	100x/0.7-1.4	0.09	0.3	oil immersion, 0.17 mm coverslip

for the beamsplitter not only to reflect excitation light between 600 - 1000 nm and to transmit PL light above these wavelengths, but also to transmit visible light between 400 - 600 nm. This is done to some extent by both of the dichroic beamsplitters available (Fig. 3.2). But since their transmittance in this wavelength region is quite low, the maximum intensity of the cold light source had to be used, especially with beam splitter #044 and highly magnifying objectives. In general, optical video images showed low contrast and were only used for laser focal alignment purposes as well as sample retrieval. All optical micrographs shown in this thesis were acquired with a standard bright-field microscope, equipped with a color video camera (Leica DMRE).

For a measurement at a certain position on a sample, it is necessary to know where the excitation light hits the sample. This laser focal alignment is done in a combination of PL measurement- and optical mode by reducing laser intensity to a minimum, removing longpass filter (d), flipping mirror (a) into the beam and switching on the cold light source (16). In this way, the focal spot of the laser can be seen in the video image. The focal point is then aligned to hit the sample approximately in the middle of the field of view of the video camera and the position is marked in the video software (WITec).

Switching between PL measurement- and optical mode requires touching the microscope which normally means that the current position on a sample is lost. This is problematic if individual nanotubes on a surface are under investigation. Optical components inside the microscope show a small misalignment for the two modes resulting in an additional change in focus. Thus, the microscope stays in the PL measurement mode once the desired position is found optically. However, due to thermal drift of the micro-

3 Experimental

scope body, the laser focus moves with time (~1 µm on a time scale of several hours), requiring additional corrections or the use of photoluminescent markers as described in Ref. [110, 111].

3.1.4 Spectrograph and Detector

For PL spectroscopy on nanotubes and especially on individual SWNTs, a highly sensitive NIR detection system is needed. Most research groups working on the PL of SWNTs currently use Indium-Gallium-Arsenide (InGaAs) detectors which show a high quantum yield for wavelengths up to 2200 nm. Our set-up employs a 300 mm spectrograph [Roper Scientific, 'Acton SpectraPro 2300i', (2)] which can be equipped with up to 3 different gratings and a 1D InGaAs photodiode array [Princeton Instruments, 'OMA-V', (1)].

Two gratings are installed in the spectrometer, one with 830 grooves/mm, blazed for 820 nm and one with 150 grooves/mm, blazed for 1200 nm. The former is used to calibrate the lasers, i.e. to correlate the position of the actuators to the output wavelength. Only the latter is typically used for the PL of nanotubes[1]. The relatively low number of grooves generates low losses and covers a spectral region of ~530 nm. Additionally, all mirrors and gratings inside the spectrometer are gold-coated.

The detector is a linear photodiode array with 1024 pixels (each 25 x 500 µm), cooled to -100 °C with liquid nitrogen. Its specified spectral range is between 950 and 1600 nm with a quantum efficiency of 75-80%. The sensitivity drops off quickly outside this range. Spectrograph and detector are calibrated using a xenon gas discharge lamp (LOT Oriel). The best achievable wavelength accuracy is on the order of ±1 pixel, corresponding to ±0.5 nm in combination with the 150 mm^{-1} grating.

3.1.5 Photoluminescence Excitation (PLE) Mapping

Part of this Ph.D. thesis encompassed the automation of photoluminescence excitation (PLE) mapping measurements. PLE mapping in the case of carbon nanotubes means that PL emission spectra are recorded for a certain range of excitation wavelengths. The excitation wavelength is varied in steps, usually in increments between 1-5 nm. For the duration of the measurement, it is important that the piezo stage and the microscope should not move such that the same sample position for all emission spectra is

---
[1] For low temperature PL measurements, the first grating was also sometimes used

3.1 Confocal Photoluminescence Microscope

guaranteed. For the excitation, a self-written 'Labview' program (National Instruments) controls LPC (24), shutter (22) and motorized actuators on the filter wheel (21) and the lasers (14, 17, 19). Spectrometer and detector are controlled using 'WinSpec' (Roper Scientific). The software displays the collected spectra and allows for external triggering of the InGaAs camera. The 'Labview' program requires start and end values of the desired excitation range as well as step size, acquisition time for each spectrum and excitation power. It then automatically scans through the excitation wavelengths, opens the shutter, adjusts the power and sends a trigger signal to the InGaAs camera to acquire a spectrum. It waits for the spectrum to be stored before closing the shutter and moving to the next wavelength. The acquired multi-spectra file has three dimensions: emission wavelength, excitation wavelength and intensity. It can be plotted as a 3D graph but for PLE maps with multiple signals, a color-coded intensity contour plot is better suited for the sake of clarity. Plots are created by importing the multi-spectra file into 'Origin' (OriginLab Corporation). Typically in mapping an individual SWNT, the maximum in emission (ideally) corresponds to the lowest optically active excitonic state (E_{11}) and the maximum in excitation to the E_{22} state. Both values allow a reliable assignment of nanotube chirality. As already mentioned, thermal drift of the microscope or of the piezo stage limits the total duration of a measurement. Typically, PLE mapping took between 10 and 30 minutes. Besides automatic control, the 'Labview' program can also be used to set the lasers to specified wavelengths and powers. This is used e. g. to show the stability of a PL signal with time. For this kind of measurement, the excitation wavelength, power and sample position stay constant and the spectra are simply collected one after the other. In a contour plot, the axis 'excitation wavelength' is then replaced by 'time' and the extent of a blinking behavior or a deterioration of the signal due to bleaching can be established (see Sec. 5.2 and 5.4.4).

3.1.6 Photoluminescence Imaging

In PLE mapping, the excitation wavelength is scanned while the sample position is held constant. For PL imaging, exactly the opposite holds. The excitation wavelength is set to a value close to E_{22} of the SWNT under investigation. The movement of the piezo stage is controlled by a program called 'PZT control' (Physik Instrumente). It uses its own macro language to define movements for all 3 axes. In this work, the piezo stage was programmed to perform raster scans only in (x, y)-direction with constant z. More

3 Experimental

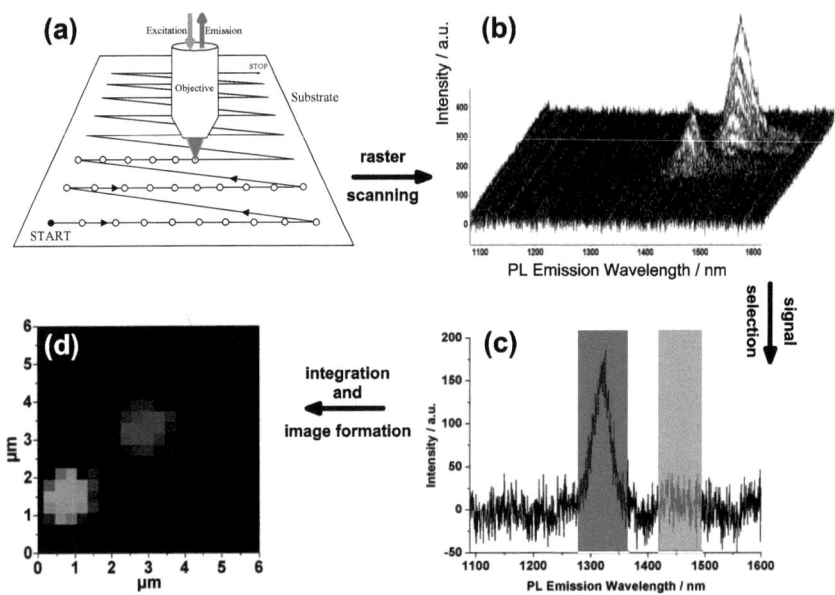

Figure 3.3: Sketch of PL Imaging technique. Pointwise raster scanning (**a**) of the sample leads to a multi spectra file (**b**). Multiple relevant signals are chosen and e.g. integrated (**c**) which results in a PL image (**d**).

complex scans of 3D objects were sometimes used, e.g. for the investigation of photonic crystals, see Ref. [139]. The macro contains start position, dimension and step size in x and y direction and acquisition time. It then initializes a pointwise scan and collects a full PL spectrum at each point, waits for the spectrum to be stored and moves a given step size distance to the next point. Figure 3.3 shows the procedure for PL imaging. The collected data is 4 dimensional: x-position, y-position, emission wavelength and intensity. For a reasonable image presentation, the spectrum has to be collapsed to a single value. This is done either by 'Spectra Analyser' or 'WITec Project' (both WITec) software. 'Spectra Analyser' is an older version which is able to integrate a spectrum over a certain wavelength range and to display the integrated value as a color code. It handles any rectangular image size. In addition, it can calculate and display (again as a color code) average, minimum, maximum and center of mass values and perform background subtraction. However, the program is quite simple and does not allow manual adjustments. It often fails to calculate correct background-subtracted images and cannot display multiple integration areas using different colors. 'WITec Project' has this capability as seen in Fig. 3.3(d). Unfortunately, 'WinSpec' multi spectra files are always displayed as quadratic images, no matter if the scanned area is rectangular or not. This 'bug' prevents the exclusive use of this program as images of long SWNTs usually exhibit high aspect ratios. PL images were predominantly acquired by the Olympus objective with a resolution of ~ 400 nm. According to the Nyquist theorem, a maximum step size of 200 nm can be employed to take full advantage of this resolution. PL images presented in this thesis were mostly recorded using this step size. Times for recording PL images ranged between half an hour and up to 12 hours, crucially depending on their x and y-dimensions. Within a time span of several hours, thermal drift is very probable. A lateral drift (x,y-direction) thereby distorts the image. This can be corrected for, but a vertical drift (z-direction) causes a loss in resolution and irretrievably blurs the image.

In addition to PL imaging, the piezo stage can be moved to any desired position within its range if a program outputs the coordinates via RS-232. This feature was used for the correlation of AFM and PL information as described in Ref. [110, 111]. AFM images were loaded into the 'Origin' software. A self-written program caused the piezo stage to move to coordinates, e.g. a nanotube, which were chosen by clicking on a position on the AFM image. If the relative shift between AFM and PL origin was determined correctly, a PLE map could be measured subsequently on the desired nanotube.

3.2 Confocal Raman Microscope

Resonant Raman measurements on carbon nanotubes are performed with a commercial confocal Raman microscope (WITec, 'CRM 200'). The set-up shown in Fig. 3.4 is very similar to that of the PL microscope, except for its optimization for Raman scattering. In the Raman measurement mode, samples are excited either by an argon ion laser (12) at 514 nm or a helium-neon (HeNe) laser (16) at 633 nm. Laser power is adjusted or shut off by a filter wheel (18) and a shutter (17). Wavelength specific, polarization-maintaining single-mode glass fibers (14 and 15) guide the light into the microscope. A holographic beamsplitter cube (6) reflects monochromatic excitation light in ($-z$)-direction, linearly polarized in x-direction. It is focused to a diffraction-limited spot using a 100x/0.9,

51

3 Experimental

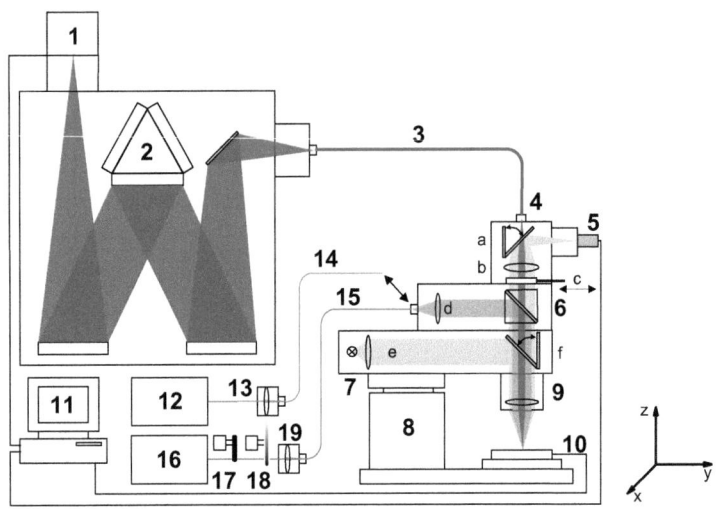

1	Si CCD Detector	13	fiber coupler (514 nm)
2	spectrograph/ triple grating turret	14	single-mode fiber (514 nm)
3	multi-mode glass fiber	15	single-mode fiber (633 nm)
4	adjustable confocal pinhole	16	helium-neon laser
5	color video camera	17	laser shutter
6	holographic bandpass beamsplitter	18	filter wheel
7	halogen lamp	19	fiber coupler (633 nm)
8	electric focusing stage	a	flip mirror
9	objective	b	tube lens
10	piezo scanning stage	c	holographic super-notch filter
11	computer	d	achromatic lens
12	Argon ion laser (514 nm)	e	Köhler illumination
		f	beamsplitter (50:50)

Figure 3.4: Sketch of the Raman microscope (WITec, 'CRM 200'). Yellow color represents visible light, laser excitation light at 514 and 633 nm and Raman scattered light is drawn in green, orange and red, respectively.

20x/0.4 or a 10x/0.25 objective [Nikon, (9)]. These objectives are parfocal, so that the specimen does not have to be refocused when the objective is changed using the objective turret. 10x and 20x objectives are normally used for coarse sample positioning in the optical mode and the 100x objective for measuring. With flip mirrors (a) and (f) moved out of the optical pathway, the Raman scattered light is focused into a multi-mode glass fiber via a tube lens (b). Residual laser light is filtered out by a holographic super-notch filter (c). Measurements in this study employed a 50 and a 100 µm fiber, which yields a lateral resolution of 300-350 nm for the wavelengths used. The glass fiber guides the light to a spectrometer [Roper Scientific, 'Acton SpectraPro 2300i', (2)], equipped with two gratings, one with 600 grooves/mm, and one with 1800 grooves/mm, both blazed for 500 nm. The detector [Roper Scientific, 'Spec-10:100B-TE', (1)] is a two dimensional silicon CCD array with 1340 x 100 pixels, thermoelectrically cooled to -45 °C. In Raman measurement mode, the 100 rows of the array are vertically binned to yield a spectrum with maximum intensity. Apart from using different fiber couplers (13, 19) and fibers (14, 15) it is necessary to exchange the whole module containing lens (d) and holographic beamsplitter cube (6), as well as holographic super-notch filter (c) and adjustable confocal pinhole (4) when changing the excitation laser line from 514 to 633 nm.

In the optical mode, laser light is shut off by shutter (17), a halogen lamp (7) is turned on and beamsplitter (f) and flip mirror (a) are put in the optical path. The super-notch filter can be removed, if necessary. Illumination is performed in a standard Köhler scheme, adumbrated by lens (e) in Fig. 3.4. A video camera (5) transfers a color image to the computer (11).

Like for the PL microscope, the sample is positioned on a combination of a piezo scanning stage (10) and a manual (x,y)-translation stage. The same kind of piezo stage, control and evaluation software is used. Therefore, Raman imaging is performed in the same way as PL imaging.

3.3 CVD Reactor

The thermal chemical vapor deposition (CVD) reactor was built for the direct growth of individual and long carbon nanotubes on substrates whose surfaces comprise predeposited, small amounts of catalyst material. Figure 3.5 shows the design of the set-up. A carbonaceous precursor and auxiliary gases flow through a 1 inch diameter quartz tube inside a horizontal, 45 cm long tube furnace (Carbolite, 'MTF 12/38/400') in which they can be heated up to a maximum temperature of 1200 °C. Available precursors are ethylene, methane, carbon monoxide, ethanol and acetylene. Hydrogen, argon and water

3 Experimental

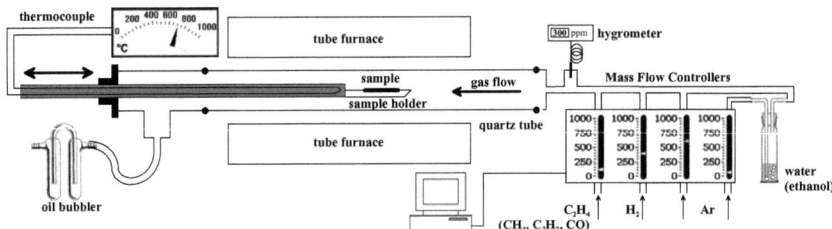

Figure 3.5: Design of the home-made thermal CVD reactor. A variety of different carbonaceous precursors and auxiliary gases can be used. Values (temperature, mass flow, etc.) and gases listed correspond to typical settings for the growth of **v**ertically **a**ligned carbon **n**anotube arrays (VANTAs, see Sec. 4.1).

vapor are employed as auxiliary gases. For the controlled mixing of all gases/vapors, four digital mass flow controllers (MFC), connected to a computer, are used (Bronkhorst, 'El-Flow'). Three of them have a flow range between 30 and 1500 sccm and one is for a flow between 0.6 and 30 sccm. A glass frit bubbler is used if ethanol is the carbon feedstock, or if small amounts of water vapor are to be added to the gas mixture. For the exact determination of ppm amounts of water vapor, a calibrated hygrometer (Michell Instruments, 'Cermet II') is placed in front of the furnace. In this case it is important to remove all traces of water (and oxygen) from the gases before a controlled amount is added. This is done by a gas-purifying cartridge (Air Liquide, 'Hydrosorb', 'Oxysorb', not shown in Fig. 3.5). The sample, which can either be a catalyst deposited on a substrate or a catalyst powder is put on a 1 x 4 cm sample holder attached to a K-type sheath thermocouple. The thermocouple measures the temperature close to the sample as opposed to the temperature gauge of the furnace. The 'Inconel' sheath of the thermocouple is sealed with O-rings which allow to quickly move the sample in and out of the hot zone. A sample holder made of molybdenum showed best resilience against high temperatures and carbon atmospheres. The gases leave the furnace via a bubbler filled with mineral oil to shut off the reaction tube against the atmosphere.

4 CVD Synthesis of Carbon Nanotubes

This chapter covers the synthesis of carbon nanotubes via the CVD set-up described in Sec. 3.3. The main focus is on vertically aligned carbon nanotube arrays and on randomly oriented, suspended SWNTs which were used to quantify the influence of external dielectric screening on optical transition energies (see Sec. 5.1), to investigate the PL at cryogenic temperatures (see Sec. 5.2) and to show experimentally the existence of 'weakly allowed' excitonic states below the lowest bright E_{11} exciton (Sec. 5.3). Additionally, the synthesis of long and partially aligned horizontal SWNTs is discussed which are spectroscopically characterized in Sec. 5.4.

4.1 Vertically Aligned Arrays of Carbon Nanotubes

The work on VANTAs was inspired by Hata *et al.*'s discovery of 'supergrowth', already mentioned in Sec. 2.5.3 [123]. Interestingly, they were able to measure PLE spectra on the as-grown 'forests' which strongly indicates that the SWNTs are not heavily bundled. Together with Raman spectroscopy of the radial breathing mode and HRTEM images, they estimated the diameters of the SWNTs to be in the range of 1 to 3 nm. We reasoned that the PL of thin as-grown SWNTs in such forests might provide a better insight into the influence (e. g. the dielectric screening) of surfactant and water molecules when compared with available dispersions of HiPco and PLV nanotubes. Therefore, the parameters given in Ref. [123] were used as starting conditions.

The best results (SWNT forest height of 2.5 mm in 10 minutes) obtained by Hata *et al.* were found for a silicon wafer substrate with thermal SiO_2 of 600 nm thickness, which was covered with magnetron-sputtered layers of 10 and 1 nm Al_2O_3 and Fe, respectively. The Al_2O_3 underlayer is supposed to increase the surface area, bind the iron particles

4 CVD Synthesis of Carbon Nanotubes

to the surface and prevent their agglomeration at higher temperatures ([140–142] and references therein).

In our approach, Si wafers covered with a 800-nm-thick layer of thermally grown SiO_2 were used. Sputtering was performed in a magnetron sputtering apparatus at the 'Physikalisches Institut' at the University of Karlsruhe, equipped with one radio frequency (RF) generator for the sputtering of insulators and three DC power supplies which are used for metals. Al_2O_3 and Fe targets (50 x 3 mm and 50 x 1 mm, respectively, Goodfellow) were sputtered with a power of 150 and 10-40 W, respectively, at an argon pressure of about $8 \cdot 10^{-3}$ mbar. Sputter rates for iron were measured by a quartz crystal microbalance. Sputter rates for Al_2O_3 had to be determined indirectly by measuring the height of a deposited layer via a high-resolution profiler (Ambios Technology, 'XP-2'), because the RF generator interfered with the microbalance. Substrates with a 10/1 nm bilayer of Al_2O_3/Fe were fabricated in this fashion.

AFM images in Fig.4.1(a) and (b) (Veeco Instruments, 'Multi Mode') show that the average surface roughness (R_a) of the bilayer is about 4 Å, as opposed to 3 Å for a plain wafer. This indicates that the sputter parameters used do not significantly increase the surface roughness. In SEM (Zeiss, 'LEO 1530', 10 kV acceleration voltage), the bilayer can be distinguished from a plain SiO_2 surface due to a difference in contrast, see Fig. 4.1(d). In the paper of Hata et al., CVD growth was carried out at ambient pressure at 750 °C for 10 minutes in a flow of 600/400/100 sccm of argon, hydrogen and ethylene, respectively and a water concentration between 20 and 500 ppm (sccm denotes **s**tandard **c**ubic **c**entimeter per **m**inute at STP). We assume that the sample was placed in a cold furnace and heated to the final temperature under a flow of 600/400 sccm argon/hydrogen, before an ethylene/water mixture was introduced. We therefore first checked the effect of such a treatment on our substrates. To prevent the catalyst from structural transformation upon slow cooling, we pulled the substrate out of the hot zone of the furnace as quickly as possible. The AFM topography image in Fig. 4.1(c) shows many small particles with an average height of ~ 5 nm which presumably are made of iron and iron oxide and have formed due to agglomeration of the sputtered Fe layer [143]. In order to increase the contrast of the particles for SEM images, a small flow of ethylene was introduced for a few seconds after the substrate had reached the target temperature. The result is seen in Fig. 4.1(e), where the area covered with the Al_2O_3/Fe bilayer has developed a 'carpet' of nanotube nucleation sites, probably made

4.1 Vertically Aligned Arrays of Carbon Nanotubes

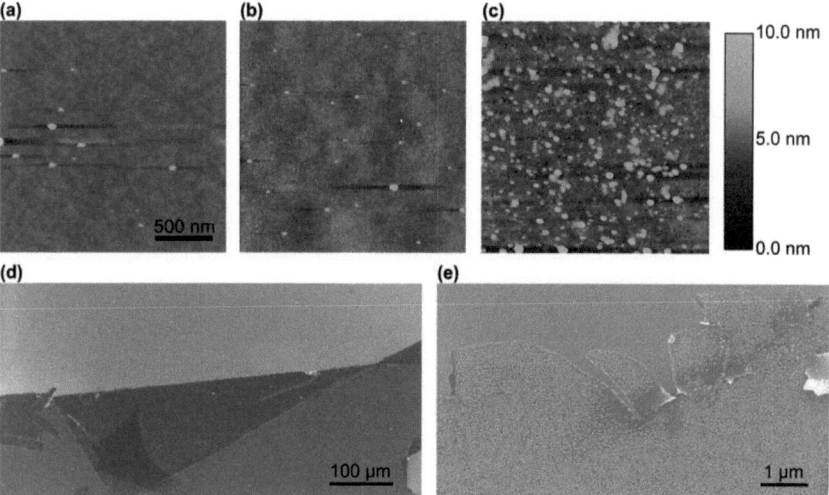

Figure 4.1: (a)-(c) AFM (2x2 µm) and (d)-(e) SEM images of substrates for VANTA growth. (a) Plain Si/SiO$_2$ wafer before sputtering. $R_a \approx 3$ Å (b) After the deposition of 10 nm Al$_2$O$_3$ and 1 nm Fe. Average surface roughness increases to ~ 4 Å. (c) After heating of (b) to 750 °C in an Ar/H$_2$ flow. Iron has agglomerated into particles with an average size (height) of ~ 5 nm. (d) Sputtered Al$_2$O$_3$/Fe bilayer (bottom) shows a strong contrast to the plain Si/SiO$_2$ surface. (e) Same as (c) plus a few seconds of ethylene flow.

of iron particles covered with a certain amount of carbon, while the bare Si/SiO$_2$ surface remains clean.

Before starting CVD growth, the temperature distribution inside the quartz tube was measured along the length of the furnace. We wanted to quantify the length of the hottest zone as well as its shift in position for a typical gas flow. Figure 4.2 shows the results for no flow (black dots) and for a 700/400 sccm flow of a Ar/H$_2$ mixture. Both maxima are close to the middle of the furnace, about 1.5 cm apart from each other. The hottest zone extends to about 5.5 − 6 cm in both cases. The temperature measured by the thermocouple in equilibrium differs by only 4 °C or 0.5 % from the value given by the furnace gauge.

Having determined the best position for the sample in the furnace, we started CVD growth experiments. Figure 4.3(a) presents a picture of an as-produced, ~ 1- mm-tall

57

4 CVD Synthesis of Carbon Nanotubes

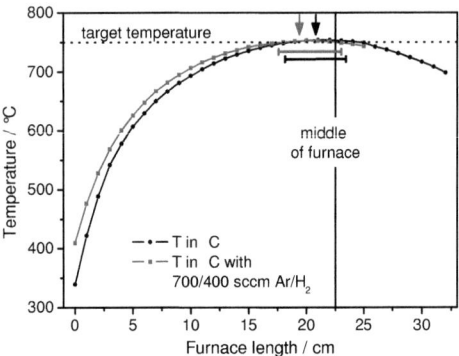

Figure 4.2: Temperature profile along the length of the furnace, without (black) and with (red) a gas flow, measured by the sheath thermocouple. The set target temperature of the furnace is 750 °C. Temperature maxima are marked with red and black arrows and are shifted by only ~1.5 cm. Constant temperature zones in both cases are 5.5-6 cm long, denoted by red and black bars.

nanotube forest after a growth time of 90 minutes. The red object next to it is the tip of a match for size comparison. SEM images (c)-(f) show different views of the VANTA and (b) is a typical TEM image displaying MWNTs with a diameter of about 5 nm and 3-4 walls. At low magnification, the nanotubes appear vertically well-aligned. Images taken on the micrometer scale, e.g. Fig. 4.3(e), reveal a loose packing of nanotubes and bundles thereof, in accordance with average densities of ca. $0.006\,\mathrm{g/cm^3}$ which were calculated from the measured heights, areas and weights of various forests. A sparse network of nanotubes or bundles suspended in different directions is characteristic for the top surface of the VANTAs [Fig. 4.3(f)]. A density of $0.006\,\mathrm{g/cm^3}$ corresponds to ~70 nm intervals between catalytically active iron particles, if the forest is approximated by a uniform array of 5- nm-thick triple-walled carbon nanotubes. This is confirmed by the AFM and SEM images of the nucleation sites in Fig. 4.1(c) and (e).

The growth parameters of this sample were chosen close to the values of Hata *et al.*, albeit with four times reduced flow rates: The chip (1 x 1 cm) with a sputtered bilayer of $10/1\,\mathrm{nm}$ Al_2O_3/Fe was heated to 750 °C in a flow of 150/100 sccm argon/hydrogen. When thermal equilibrium between sample and furnace was established, 25 sccm ethylene together with about 300 ppm water vapor were added to the argon/hydrogen flow. After 90 minutes, the flow of ethylene, water and hydrogen were stopped and the sample was slowly pulled out of the hot zone until it cooled to about 300 °C. At that temperature, nanotubes can be safely exposed to air without being oxidized.

Although the height of our nanotube forests is impressive, the growth rate is still ~20 times lower than that given in the paper by Hata. The TEM image in Fig. 4.3(b) also

4.1 Vertically Aligned Arrays of Carbon Nanotubes

Figure 4.3: VANTA grown with water-assisted CVD. (a) Picture of a ~1 mm-tall nanotube forest. The tip of a match on the right is for size reference. (b) Typical TEM image of the tubes. (c)-(e) SEM side view at different magnifications and (f), top view.

suggests that our VANTAs consist primarily of MWNTs rather than showing a high SWNT content. In order to clarify the rate aspect, we conducted a series of experiments varying the gas composition, the flow rates, the temperature and the water vapor content for both the heat-up and the growth phase. However, the number of parameters is too large to vary them all in a systematic manner, especially if they are mutually dependent. In a first approximation, we assumed independent parameters and varied only one at a time. The growth height was measured with an optical microscope (Leica DMRE) by focusing on the surface of both the Si/SiO_2 chip and the top of the forest. The average height was taken as the average from 4 – 6 measurements. In addition, Raman and PL spectra as well as some TEM images and diameter histograms via AFM were recorded

4 CVD Synthesis of Carbon Nanotubes

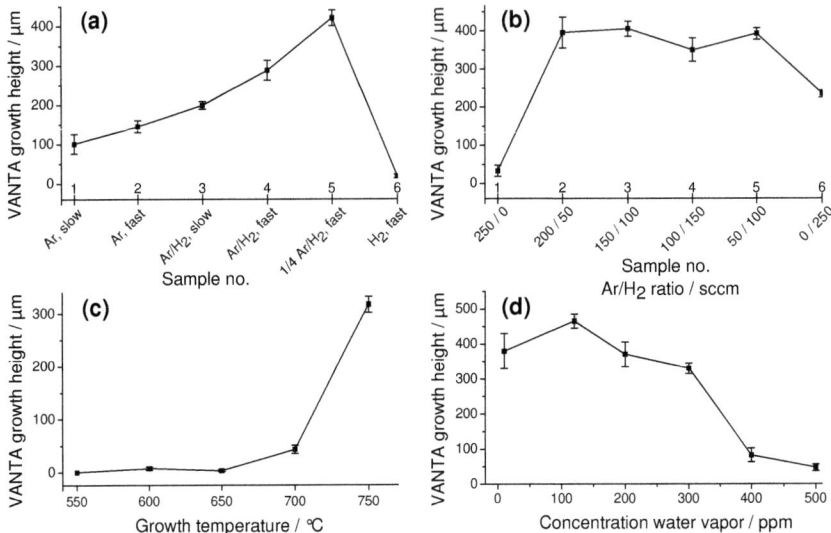

Figure 4.4: VANTA height for different parameters after a growth time of 20 minutes. **(a)** Effect of fast (sample no. 2,4,5 and 6) vs. slow (sample no. 1 and 3) heating and gas composition in the heat-up phase. See text for details. **(b)** Effect of Ar/H$_2$ composition in the growth phase. **(c)** Influence of temperature and **(d)** content of water vapor on the height.

(*vide infra*) to clarify the existence of MWNTs or SWNTs.

Some typical results are depicted in Fig. 4.4. All four plots show the forest height after 20 minutes growth time. In Fig. 4.4(a), we considered the effect of the gas composition in the heat-up phase as well as the heating rate. 'Slow' means that the sample is inside the furnace and heats up together with it, as opposed to 'fast', where the sample is pushed into the already hot furnace. For the latter, growth, i.e. the introduction of ethylene is started when the thermocouple indicates a temperature of $\sim 10\,°C$ below the target temperature. In the former case, growth is started when thermal equilibrium is reached. For sample no. 1-4 and 6 in Fig. 4.4(a), the gas composition in the growth phase was always 600/400/100 sccm Ar/H$_2$/C$_2$H$_4$. In the heat-up phase, it was only 600 sccm Ar for sample no. 1 and 2, 600/400 sccm Ar/H$_2$ for 3 and 4 and 400 sccm H$_2$ for 6 (sample no. 5: below). The data suggests that rapidly heating the sample yields higher

4.1 Vertically Aligned Arrays of Carbon Nanotubes

VANTAs[1] and that neither argon nor hydrogen alone are good choices. The iron layer is covered with iron oxide at the beginning because the chips are stored under atmospheric conditions. During heat-up, this layer gets partially reduced by hydrogen. No reduction (only argon) or a complete reduction (pure hydrogen) result in lower growth rates. The best choice seems to be a mixture of both. Another positive effect was observed upon the reduction of all the flow rates by a factor of 4, see sample 5 (150/100 sccm Ar/H_2 and 150/100/25 sccm $Ar/H_2/C_2H_4$, during heat-up and growth phase, respectively).

Figure 4.4(b) shows the effect of varying the Ar/H_2 gas composition during the growth phase. The total flow rate (250 sccm), the flow rate of ethylene (25 sccm) and the gas composition during the fast heat-up [150/100 Ar/H_2 as in sample 5 of Fig. 4.4(a)] was kept constant. Again, the growth rates for no hydrogen (sample 1) and hydrogen only (sample 6) are not as high as for a mixture of both, whereas the exact composition of the mixture seems to be of minor importance.

Figure 4.4(c) shows the influence of different growth temperatures. The total VANTA height seems to increase exponentially, starting at about 650 °C. This Arrhenius type behavior suggest the existence of an activation barrier, most likely for the decomposition of ethylene. Between 750 and 800 °C, the decomposition of ethylene resulted in a polymerization/cyclization reaction, filling the exhaust part of the quartz tube with white smoke and liquid/oily products, without the formation of nanotubes.

The water vapor concentration in the growth phase was changed in Fig. 4.4(d) between 10 and 500 ppm. Maximum growth height was achieved for a water vapor concentration of 120 ppm. A further increase of the water content led to a monotonous height decrease. However, the effect of water is not very large. The average height with (almost) no water was 380 ± 35 µm and 465 ± 20 µm with a water vapor concentration of 120 ppm, which is an increase of about 20 %.[2] Thus, a fast-heating process at 750 °C along with 150/100 sccm Ar/H_2 in the heat-up and 150/100/25 sccm Ar/H_2/ethylene plus 120 ppm water vapor in the growth phase yielded a maximum VANTA height of 465 ± 20 µm after 20 minutes or about 1.3 mm after 90 minutes.

We also varied the height of the iron layer between 0.2 and 1 nm. The effect of this can best be studied, if there is an iron height gradient on a single chip. For this purpose,

---
[1] It also increased the number of iron particles which were carried along with the growing nanotubes

[2] In Ref. [144] we reported a height increase by a factor of 10 for 300 ppm water vapor. This preliminary result was reached with uncalibrated rotameters and could not be reproduced after the modification to digital mass flow controllers.

4 CVD Synthesis of Carbon Nanotubes

Figure 4.5: 6 x 1 cm chip covered with vertically aligned nanotubes. The chip was prepared with 10 nm of Al_2O_3 and an iron gradient from 1 to 0.2 nm. Numbers were written with a pen on the chip before sputtering and removed with acetone before CVD growth. The iron catalyst layer is thickest around 1 and decreases in height towards 11.

we used the nonuniform sputter rate distribution at the edge of the sample holder inside the sputter apparatus. A 6 x 1 cm chip was prepared and marked with numbers from 1 to 11. As usual, 10 nm of Al_2O_3 were deposited first. Close to 1, the height of the iron layer was 1 nm, and decreased to about 0.2 nm close to 11. Figure 4.5 shows the result after 90 minutes of CVD growth. Close to 1, the height of the VANTA is again more than 1 mm and slowly decreases towards 11. However, Raman spectra recorded along the length of the chip do not indicate a change from MWNTs to SWNTs. It seems that the height of the iron layer in this range is of minor importance, because upon heating, iron agglomerates to particles of the same size and (presumably) different particle densities. Thus, the VANTAs at the end with the thin iron layer should grow less dense. However, this was considered of minor importance regarding the growth of thin SWNTs and therefore not quantified in this work.

The amount of iron carried along with the growing nanotubes was checked by reusing the substrates. VANTAs are easily removed from the substrate (e.g. with a scalpel), in fact, they almost come off by themselves. We found that chips that were used with fast-heating hardly grew any VANTAs the second time. When slow-heating was applied, they could be reused at least once, although the maximum achievable height decreased substantially. Substrates with an additional layer of only 1 nm Al_2O_3 on top of the iron layer however could be reused more than 5 times, without a (noticeable) loss of growth rate. The additional layer of alumina seems to increase adhesion of the iron particles. Yet, it also decreases the maximum achievable height to a few tens of micrometers.

In order to measure a diameter distribution of our VANTAs, we performed a sample preparation for AFM analysis which has been described elsewhere [145]. Briefly, CVD-

4.1 Vertically Aligned Arrays of Carbon Nanotubes

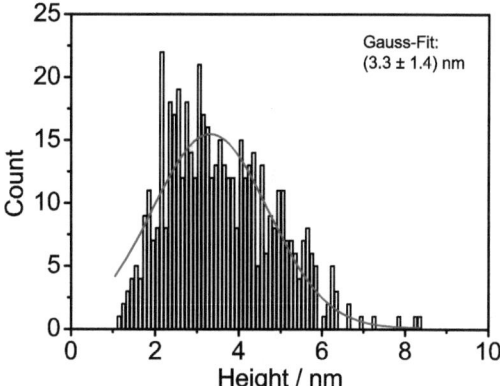

Figure 4.6: Typical histogram of diameters (heights above the surface) for individualized CVD-grown nanotubes spin-coated onto a silicon surface from a sonicated aqueous dispersion and measured by AFM imaging (see the text for details). The red curve denotes a Gaussian fit with a mean diameter of 3.3 ± 1.4 nm. From Ref. [144].

grown VANTAs were removed from the substrate and dispersed in water plus 1 wt % of sodium cholate with an ultrasonic tip disperser. Nanotubes were then spin-coated onto a silicon wafer, rinsed with acetone, and dried at room temperature. AFM images were recorded in intermittent contact mode and analyzed with the 'SIMAGIS' software (Smart Imaging Techn.). Figure 4.6 shows a typical histogram of diameters (heights above the surface) for our VANTAs. A variation of the parameters mentioned above did not significantly change this distribution. The maximum remained between 3 and 4 nm which is consitent with the average size distribution of the iron particles in Fig. 4.1(c) and which is also close to the value of 3 nm reported by Hata et al.

The AFM histogram in Fig. 4.6 suggests that only approximately 8 % of the nanotubes generated have diameters between 1 and 2 nm (and are correspondingly SWNTs). Accordingly, the dispersion prepared from the CVD tubes shows very weak PL (limited to diameters ≤ 1.4 nm). A Raman spectrum of the dispersion at 785 nm excitation (Kaiser Optical Systems Inc., 'Holospec') is shown in Fig.4.7(a). It displays a rather high D-mode and a single G-mode, typical for defective MWNTs. There is also a rather untypical, broad peak in the RBM region around 170 cm^{-1} (equivalent to SWNTs with diameters around 1.4 nm). Raman and PL microscopy on the as-grown VANTAs was more informative. Some characteristic spectra are presented in Fig. 4.7(b)-(d). Figure 4.7(b) shows a Raman spectrum from the top surface of the array with a low magnification objective. Again, D- and G-mode but no RBMs are present, irrespective of which part of the sample was probed. With a high magnification objective [100x/0.9,

4 CVD Synthesis of Carbon Nanotubes

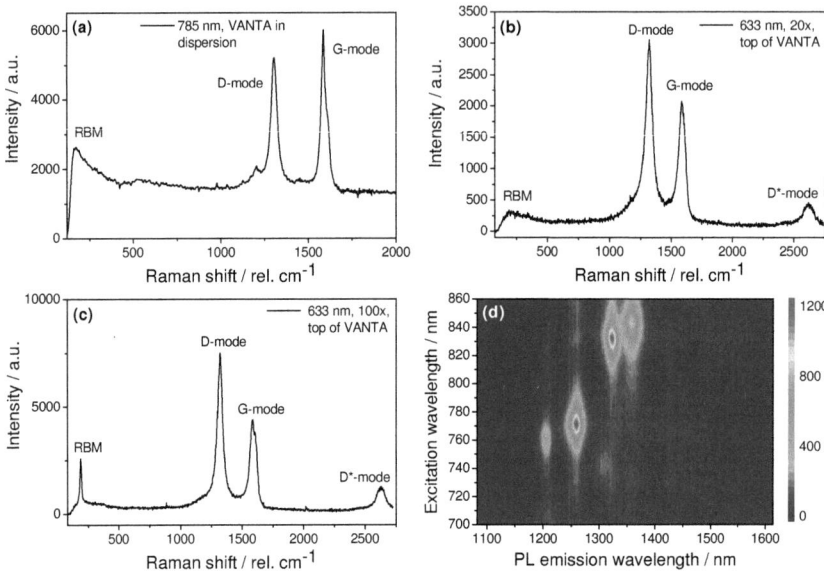

Figure 4.7: (a)-(c) Typical Raman and (d) PLE spectra obtained for VANTAs generated in this work. (a) Raman spectrum of VANTAs dispersed in D_2O/1 wt % SDBS and excited with 785 nm shows a broad signal in the RBM region as well as D- and G-modes. (b) and (c) are Raman spectra from the top of as-grown VANTAs at 633 nm excitation with 20x/0.4 and 100x/0.9 objectives, respectively. At higher magnification, RBM signals of individual SWNTs can be observed (here at $\sim 188\,\text{cm}^{-1}$, corresponding to a diameter of approx. 1.27 nm). (d) PLE spectrum from the top of as-grown VANTAs showing 4 distinct chiralities.

Fig. 4.7(c)], the spectrum strongly depended on the position and different RBMs were observed. Similarly, PL was only observed at high magnifications. When the signal was carefully optimized by adjusting the position with the piezo stage, PL contour maps of individual or small ensembles of SWNTs could be recorded, as shown in Fig. 4.7(d). We attribute the above observations to a small fraction of air-suspended SWNTs extending beyond the top surface of the forests. The forests themselves mainly consist of MWNTs. We also found PL signals of individualized SWNTs on the bottom and in the middle of the VANTAs albeit with somewhat weaker and broader signals. Because SWNTs exist only as a minor fraction, and the density of the forest is sufficiently low, bundling and therefore quenching of PL is efficiently prevented. From the SEM images we estimate that suspended sections can span distances of hundreds of nanometers. The individual SWNTs detected by PL microscopy on top of the forests are likely suspended in a similar fashion.

In conclusion, we have successfully grown vertically aligned carbon nanotube arrays with heights of more than 1 mm from a sputtered alumina/iron catalyst bilayer. We have explored the influence of different parameters like the heating rate, gas composition, growth temperature and water vapor content. These parameters could be optimized to yield a maximum growth rate, but the average diameter of the tubes remained almost constant between 3 and 4 nm. Our CVD arrays consist of (typically 3-4 walled) MWNTs with an average distance of ~ 70 nm from each other. About 8% of all tubes are SWNTs with a diameter below 2 nm.[3] The small content of SWNTs in our materials is in contrast to the results of Hata *et al.* who achieved substantially faster growth (about 20 times faster) of up to 2.5- mm-high forests of ~ 1-3 −nm-diameter nanotubes using a similar CVD procedure [123]. We could not reproduce the water-assisted 'supergrowth' although the addition of 120 ppm of water vapor increased the maximum growth height by about 20 % compared to the growth with 10 ppm of water vapor. We believe that differences in catalyst preparation are responsible for discrepancies in the material composition and growth rate. However, our VANTAs are ideally suited for spectroscopy on individual SWNTs (see Ch. 5).

4.2 Suspended and Horizontally Aligned Arrays of Carbon Nanotubes

Besides vertically aligned carbon nanotube arrays employing ethylene and a sputtered catalyst, we also realized other growth morphologies using catalysts prepared by wet chemistry methods and different carbon feedstock gases, mainly mixtures of ethanol, hydrogen and argon ('ethanol CVD') carbon monoxide and hydrogen ('CO-CVD') and methane, argon and hydrogen ('methane CVD'). All led to tubes air-suspended between short cracks (1-2 µm) of the catalyst material or to long ($\sim$ several hundreds of microm-

---

[3]$\sim$2% have diameters below 1.4 nm which corresponds to the upper limit of our detection range.

eters) nanotubes resting on the substrate and being partially aligned with the gas flow.

The catalyst for both growth morphologies was prepared from 15 mg of highly dispersed pyrogenic alumina (Degussa, 'AEROXIDE® Alu C'), 0.05 mmol of iron nitrate, $Fe(NO_3)_3 \cdot 9\, H_2O$ and 0.015 mmol of molybdenyl(VI)acetylacetonate, $MoO_2(acac)_2$, dispersed in 15 mL of methanol or isopropanol. The composition of the catalyst is based on that used by Kong et al. [127]. Prior to use, the mixture was agitated by an ultrasonic tip disperser for 5 minutes. Silicon or sapphire chips (CrysTec) were partly covered with the dispersion either by dip-coating or by dripping a small amount onto the surface, thereby creating a pad of dried catalyst material. After drying in air, nanotubes were grown either using ethanol, carbon monoxide or methane as feedstock.

Ethanol CVD The sample was placed in the middle of the cold furnace and heated to 900 °C in a prefabricated mixture of argon and hydrogen (Air Liquide, containing 5 % H_2) at a flow of 800 sccm. This calcination procedure lasted between 15 and 30 minutes and is supposed to decompose all organic compounds and to partially reduce the catalyst ions to the elementar metals. Subsequently, the complete gas flow was guided through an ethanol bubbler prior to entering the furnace. Provided that ethanol vapor reaches saturation, its concentration in the gas flow is about $110\,g/m^3$ corresponding to an ethanol vapor flow of 40 sccm at 20 °C (room temperature, RT). The ethanol bubbler could be heated to about 50 °C corresponding to a maximum ethanol vapor concentration of $500\,g/m^3$ or a flow of 200 sccm. In most growth experiments we wanted to grow low densities of tubes, and the bubbler was therefore kept at RT. After 20-30 minutes growth time, the ethanol bubbler was taken out of the gas flow and the sample cooled down at ~ 100 sccm of Ar/H_2.

Figure 4.8 shows different SEM/TEM images and a Raman spectrum of carbon nanotubes synthesized with ethanol CVD. A characteristic 'island' or pad of catalyst material with crisscrossing cracks is depicted in Fig. 4.8(a). Nanotubes are frequently found to cross such trenches, as in Fig. 4.8(b), (c) and (d). The TEM image in the inset of image 4.8(a) demonstrates that the sample is made up of SWNTs with diameters of around 2 nm and bundles thereof. This is confirmed by a bulk Raman spectrum of the RBM region in the inset of Fig. 4.8(d). Diameters range between 1.1 and 2.3 nm. Figure 4.8(e) shows a long nanotube in direct contact with the Si/SiO_2 substrate. The contrast here is not as good as for air-suspended tubes. Loop structures like in the inset of Fig. 4.8(e) are also found frequently. Spectroscopic and AFM characterization of SWNTs made by

4.2 Suspended and Horizontally Aligned Arrays of Carbon Nanotubes

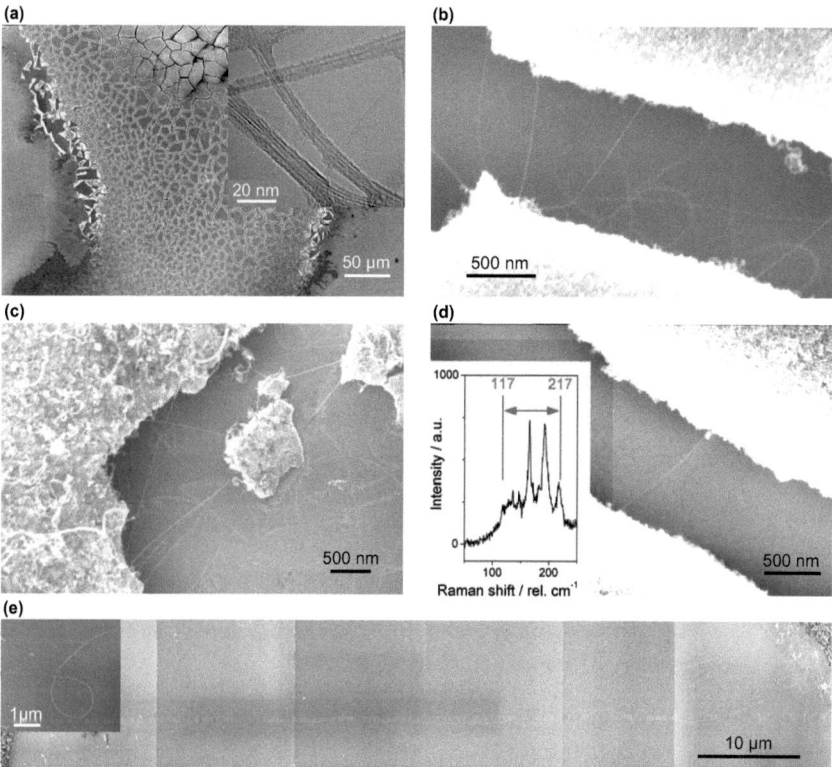

Figure 4.8: SEM images of carbon nanotubes synthesized with ethanol CVD. **(a)** Catalyst 'island' on top of a Si/SiO$_2$ chip. The inset of **(a)** depicts a TEM image of SWNTs and bundles of SWNTs with diameters of ∼2 nm. **(b)**, **(c)**, **(d)** SWNTs crossing the trenches shown in **(a)**. The bulk Raman spectrum (633 nm) of the RBM region in the inset of **(d)** indicates diameters between 1.1 and 2.3 nm. **(e)**, inset of **(e)** Individual SWNTs on the Si/SiO$_2$ substrate.

ethanol CVD is part of Ch. 5.

Carbon monoxide CVD The same catalyst dispersion was used as for ethanol CVD. In order to obtain a sharp separation between bare surface and surface covered with catalyst, part of the chip was covered by a scotch tape before the catalyst dispersion was dripped on it. After drying, the tape was removed. The sample was pushed into the

67

4 CVD Synthesis of Carbon Nanotubes

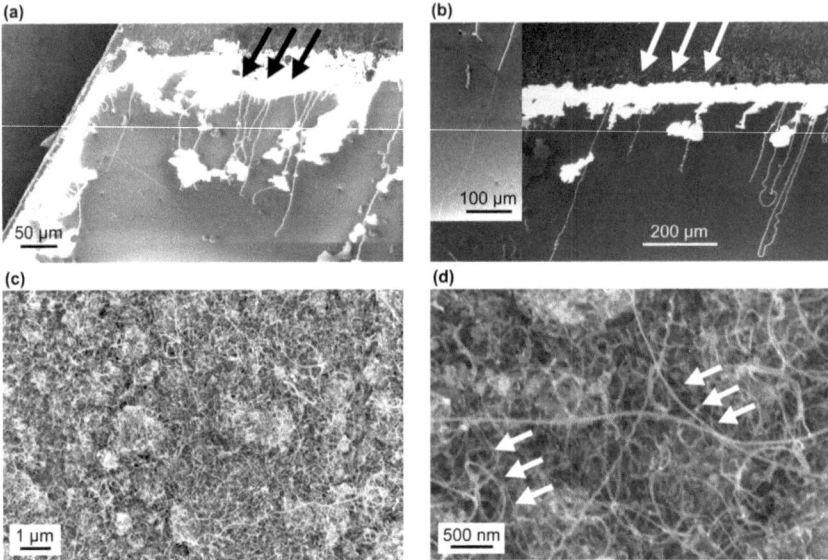

Figure 4.9: SEM images of carbon nanotubes synthesized with CO-CVD on sapphire. **(a)**, **(b)** Low magnification images, taken at 3 kV acceleration voltage showing long and partially aligned nanotubes, some with loops and U-bends. The gas flow direction is indicated by the arrows. The inset in (b) depicts a tube extending from the top right to the bottom left of the image, exhibiting a cut. This cut separates the lower part of the tube from the catalyst pad which significantly reduces contrast (see text for details). **(c)**, **(d)** Images from the top of the catalyst pad, taken at 6 kV acceleration voltage. A dense network of nanotubes with different diameters is visible. Thin and suspended SWNTs [indicated by arrows in **(d)**] can be found, as well.

900- °C-hot furnace at a flow of 800 sccm carbon monoxide and 200 sccm hydrogen. To avoid high CO concentrations in the hood, the gas mixture was combusted upon exiting the oil bubbler. After about 30 minutes of growth, the sample was slowly pulled out and the gas flow was switched to 1000 sccm argon. The growth time was taken from the moment the sample was pushed into the furnace until the flow of CO/H_2 was replaced by argon. This fast-heating CVD process is based on that of Liu and coworkers [131]. However, their catalyst was made of monodispersed Fe/Mo nanoparticles and their CVD

4.2 Suspended and Horizontally Aligned Arrays of Carbon Nanotubes

set-up consists of two furnaces. Using this procedure, they were able to grow individual, aligned SWNTs with a length of up to 2 mm and diameters between 1 and 2.5 nm. Figure 4.9(a) and (b) shows SEM images (at 3 kV acceleration voltage) of nanotubes grown on sapphire substrates via our CVD process. They are several hundreds of micrometer long and partially aligned. Some loops and 'U'-bends are also visible. Interestingly, the insulating substrate allows for the detection of tubes that are connected to the catalyst area at very low magnifications, i. e. a big field of view. Only when a tube is disconnected from the catalyst pad, does it appear darker [see inset in Fig. 4.9(b)]. These observations are consistent with what has been described in the literature, although different models explaining this effect are proposed [146–148].

According to Zhang et al. [148], insulating substrates like SiO_2 or sapphire are either positively or negatively charged at low (from 1 to 5 kV) or high (more than 5 kV) primary electron energies, respectively. When the surface of the insulating substrate is positively charged, electron accumulation on the SWNTs results in a larger negative potential relative to the substrate, which enhances the secondary electron emission from the SWNTs, and thus yields a brighter contrast. The positive potential of the scanned area can induce electrons to flow in from an electron supply. A metallic pad (or the grounded catalyst pad in our case) can serve as an electron supply source, making the contacted SWNTs (and also, within the electron diffusion length, its surroundings) more negatively charged and thus appearing brighter and thicker than the isolated ones. The same explanation probably holds for the brightness of the catalyst film edge and other particles connected to the catalyst pad via the nanotubes.

High acceleration voltages (typically 10 kV in our case) reverse the contrast between substrate and SWNT, because the substrate is now negatively charged compared to the tube and emits more secondary electrons. Since both the SWNTs and the substrate are negatively charged, electron diffusion from the former to the latter is suppressed, yielding images of SWNTs with diameters smaller than that observed on a positively charged substrate [see e. g. Fig. 4.10(a)]. Unless otherwise indicated, all of our SEM images, especially for Si/SiO_2 substrates, are taken at 10 kV acceleration voltage, because this facilitated focusing.

It has been speculated that metallic and semiconducting SWNTs could be distinguished from their SEM images, but this was not experimentally confirmed [146–148]. According to the explanation given above, the tubes with high contrast could be metallic

4 CVD Synthesis of Carbon Nanotubes

whereas the semiconducting SWNTs are dark and cannot be seen at this magnification. However, we did not find more tubes when increasing the magnification or using different scanning techniques like AFM or µ-Raman spectroscopy (see Sec. 5.4.2). Thus, we believe that the high contrast of the tubes does not reflect their metallic or semiconducting character.

SEM images in Fig. 4.9(b) and (d) are taken from the top of the catalyst area. It is covered with nanotubes of various diameters, some being partially air-suspended. Most tubes here are probably MWNTs but some thin tubes, possibly SWNTs (highlighted by white arrows), are present, as well. This was also shown by µ-PLE spectroscopy (Sec. 5.2 and 5.3).

Methane CVD In a few CVD experiments, we used methane as feedstock gas. The catalyst was either prepared by sputtering layers of 10 and 1 nm of alumina and iron on a silicon chip like previously done for ethylene CVD, or by dip-coating a substrate, utilizing the same dispersion used for the growth with carbon monoxide or ethanol. We again employed the fast-heating process by pushing the sample at a flow of 120 sccm argon and 64 sccm hydrogen into the 900-°C-hot furnace. When the thermocouple indicated a sample temperature of about 890 °C, 500 sccm methane was added to the gas flow. After a growth time of typically 30 minutes, the methane and the hydrogen flow were stopped and the sample was pulled out of the hot zone until it reached a temperature of about 300 °C. The quartz tube was then opened and the sample removed.

Nanotubes resulting from methane CVD are presented in Fig. 4.10. Figure 4.10(a) and (b) show individual tubes and a dense 'mat' grown from the dispersion and sputtered catalyst, respectively. The reason for the low contrast of Fig. 4.10(a) was already discussed above. Fig. 4.10(d) is a Raman spectrum of the nanotube mat taken at 633 nm excitation. It verifies the existence of SWNTs with a small amount of defects and diameters between 1.1 and 2.1 nm. Interestingly, the mat is only a few micrometers high, as opposed to the VANTAs in Sec. 4.1 and mostly consists of SWNTs instead of MWNTs. The PLE contour map in Fig. 4.10(d) was recorded from a single $(10, 9)$ SWNT resting on the SiO_2 surface. The broad and strong signal between 1050 and 1250 nm is background PL from the silicon substrate. The assignment of the chiral indices is possible due to Ref. [144] and will be discussed in Sec. 5.1.

Methane CVD from Li *et al.* In a cooperation with the group of Prof. Yan Li from the University of Beijing, we were also able to use their ultralong horizontally

4.2 Suspended and Horizontally Aligned Arrays of Carbon Nanotubes

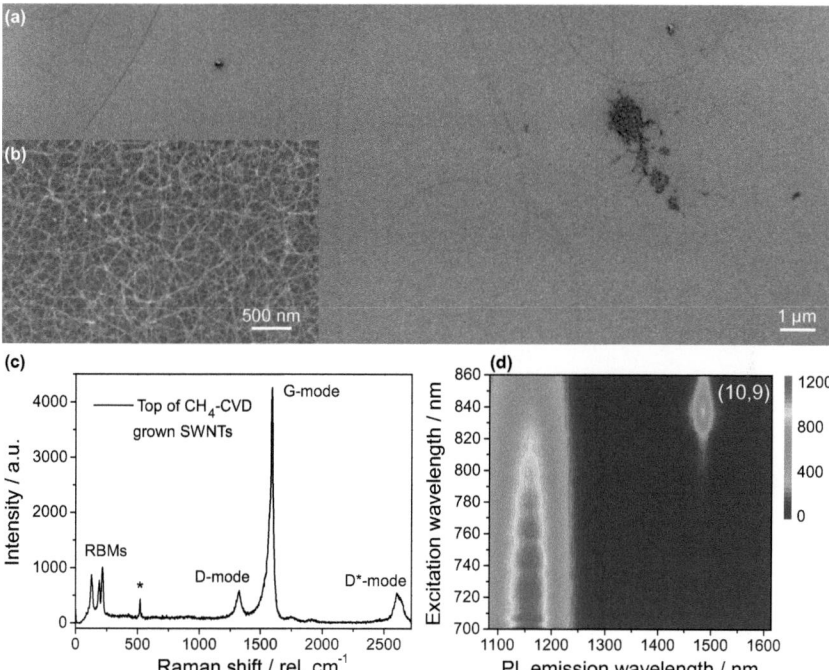

Figure 4.10: SEM images, Raman and PL spectra of carbon nanotubes synthesized with methane CVD. (a) Loop structures and straight individual tubes on the SiO_2 substrate, grown using a dip-coated catalyst. SEM acceleration voltage 10 kV. (b) 'Mat' of SWNTs grown by using a sputtered bilayer of $10/1$ nm Al_2O_3/Fe. (c) Raman spectrum (633 nm excitation) recorded with a 100x objective on the top surface of a SWNT mat. Diameters calculated from RBM frequencies range between 1.1 and 2.1 nm. The asterisk denotes a background Raman signal from the Si substrate at $521\,cm^{-1}$. (d) PLE contour map of an individual $(10,9)$ SWNT in direct contact with the SiO_2 surface. The broad signal covering most of the excitation range is due to the PL of silicon, which exhibits a band gap of $\sim 1.12\,eV$ or 1100 nm.

aligned SWNT arrays, prepared from methane CVD. The exact procedure is described in Ref. [136]. Briefly, a diluted Fe-Mo nanoparticle solution (average diameter 6.1 nm) was applied to the front edge of a piece of silicon wafer using a syringe. The substrate was placed into a furnace and heated to 900 °C in Ar atmosphere and treated with 100 sccm

4 CVD Synthesis of Carbon Nanotubes

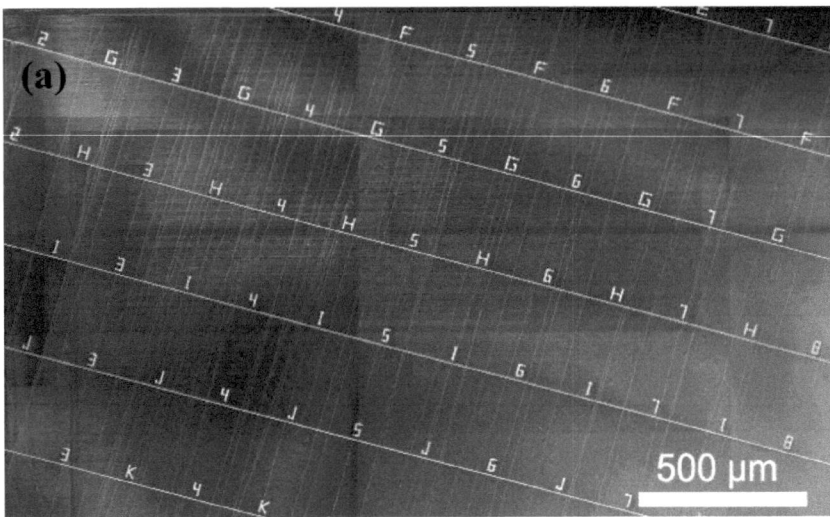

Figure 4.11: SEM image of horizontally aligned and ultralong nanotubes grown on a lithographically prepatterned Si substrate. The pattern includes etched trenches and markers. Taken from Ref. [136].

H_2 for 20 minutes. Subsequently, the temperature is increased to 970 °C and nanotube growth is performed at 0.5 sccm CH_4 and 1.0 sccm H_2 for 30 minutes. Afterwards, the sample is cooled down at about 10 sccm Ar.

Li et al. observe that the extremely low flow rate of the gas is best suited for the growth of centimeter-long SWNTs. The diameter distribution of nanotubes measured by AFM is 0.9 − 3.3 nm with a mean value of 2.0 nm. In addition, they are able to prepattern a silicon wafer with trenches and markers by photolithography and to let the tubes grow over them (see Fig. 4.11). The markers enabled us to find the same SWNT with SEM, AFM and the PL microscope. The alignment and the trenches greatly simplified the search for a nanotube showing PL. A large SWNT length allowed us to perform AFM manipulation of individual tubes and to observes changes in the PL spectra. This is elucidated in Sec. 5.4.4 .

5 Spectroscopic Characterization of Individual Carbon Nanotubes

In the following 4 sections the CVD-grown SWNTs of the previous chapter are used as model systems to study excitons in carbon nanotubes. First, the influence of a dielectric medium on the optical transition energies of SWNTs will be investigated [9]. Second, we compare CVD-grown SWNTs to surfactant-coated nanotubes at temperatures down to 4 K [10]. Third, the observation of deep excitonic states below the optically active exciton is discussed [11]. The PL of ultralong SWNTs and the effect of uniaxial and torsional strain introduced by AFM manipulation of individual tubes is presented in section four [12]. The fifth section introduces a new method for observing and counting individual nanotubes in dispersion [13].

5.1 Influence of External Dielectric Screening on Optical Transition Energies

5.1.1 Motivation

As mentioned in Sec. 2.4.6, PL [2] and Raman studies [88, 90, 101, 102] of SWNTs have led to the assignment of characteristic optical transition energies E_{ii} to specific chiralities, (n_1, n_2). The assignment of E_{ii} vs. (n_1, n_2) was primarily derived from measurements of micelle-coated SWNTs in water-surfactant dispersions, where bundles can be efficiently exfoliated into individual nanotubes. The tubes are thus free from inhomogeneous intertube interactions which practically completely quench the PL of semiconducting SWNTs. In contrast to this, most theoretical models consider nanotubes without environmental effects. It is therefore important to quantify the effect of a specific environment on the electronic structure and optical transition energies of individual SWNTs, relative to the

5 Spectroscopic Characterization of Individual Carbon Nanotubes

'ideal' case of nanotubes in vacuum. As PL originates from the radiative decay of excitons in semiconducting SWNTs, this means examining the influence of the environment on exciton binding and formation energies.

Because most of the electric field of an exciton in a 1D system such as SWNTs is located outside the nanotube, dielectric environments are expected to efficiently screen Coulomb interactions of electronic excitations [66, 149, 150]. Thus, binding energies (E_b) of excitons in nanotubes could in principle be 'tuned' by changing the dielectric constant (also called permittivity) ϵ_r of the environment, from strongest binding at $\epsilon_r \approx 1$ up to small E_b and thermal dissociation of the exciton at sufficiently large ϵ_r. Note the difference between the binding energy of an exciton, E_b and the optical transition energy E_{ii} measured in the experiment: E_{ii} is the energy difference between the ground state and the lowest bright exciton observed in experiment, A_2 or A_{2u} in chiral or zig-zag SWNTs, respectively. E_b is the energy difference between the lowest excitonic state (no matter if optically allowed or not, but the difference is negligible for typical exciton binding energies in nanotubes) and the 'series limit', i.e. when electron and hole are completely separated from each other, corresponding to $\nu = \infty$ in Fig. 2.8(c) and (d).

So far, only few measurements of optical transition energies of individual SWNTs partially suspended in air ($\epsilon_r \approx 1$), i.e. free of surfactant and other sidewall interactions, have been reported, with conflicting results [151–154]. These nanotubes were typically synthesized by catalytic CVD over prepatterned substrates.

Lefebvre et al. observed blueshifts of $\Delta E_{11} = +(19\text{-}37)$ meV and $\Delta E_{22} = +(7\text{-}28)$ meV of PLE peaks measured for several individual nanotubes grown between pillars etched in silicon relative to a 'standard' aqueous dispersion of SWNTs in SDS [151]. Yin et al. applied RRS to determine resonance energies of air-suspended SWNTs grown over trenches in silicon [152]. They contradict the assignment of Lefebvre et al. in favor of a $\sim 70\text{-}90$ meV *decrease* of E_{22} for air-suspended semiconducting nanotubes compared to the aqueous dispersion. Ohno and coworkers measured PL of SWNTs grown over a groove pattern etched in quartz [153] and observed PLE shifts on average similar to those reported by Lefebvre et al. However, as opposed to the other studies, they found a correlation of ΔE_{11} and ΔE_{22} with nanotube chirality and the S1/S2 family. Finally, Okazaki et al. were able to detect PL directly from as-prepared SWNT material synthesized by gas-phase pyrolysis [154]. Due to the weak signals, only a few (n_1, n_2) species were identified with peaks blueshifted by $\Delta E_{11} = +(51\text{-}62)$ meV and

5.1 Influence of External Dielectric Screening on Optical Transition Energies

$\Delta E_{22} = +(57\text{-}71)\,\text{meV}$, referenced again to an aqueous dispersion of SWNTs.

To resolve these discrepancies, we have carried out systematic µ-PLE spectroscopy studies using a different sample configuration, namely SWNTs occurring as a minor fraction in VANTAs grown by thermal CVD as described in Sec. 4.1. E_{11} and E_{22} energies could be measured not only in air or vacuum, but also—after careful and slow immersion—in non-volatile liquid environments. We found uniform blueshifts of E_{11} and E_{22} energies by $+(40\text{-}56)\,\text{meV}$ and $+(24\text{-}48)\,\text{meV}$, respectively and were able to assign those values to 19 different (n_1, n_2) nanotube species suspended in air or vacuum. Below, we also discuss energy shifts caused by immersion in organic liquids like 1-methylnaphthalene and paraffin oil. The differences in transition energies are compared to theoretical predictions for dielectrically screened excitons in SWNTs.

5.1.2 Sample Preparation and Measurements

The PLE measurements were performed at ambient temperature on VANTAs with a height of typically $\sim 0.2\,\text{mm}$ grown on Si and sapphire wafers. A (x, y)-translation stage and a piezo stage were used to localize luminescent SWNTs under the microscope (either in air, in an optical vacuum cell, or immersed in a thin solvent layer). For a luminescent SWNT suspended in air, a few mW of focused laser power could lead to a significant heating and consequently to small redshifts of optical transitions (PL peaks), in accordance with Raman results of Cronin et al. [155]. This was particularly true for SWNTs suspended in vacuum. Therefore, low laser powers of typically 0.1-0.5 mW, focused to power densities less than $\sim 5 \cdot 10^3\,\text{W/cm}^2$ (and correspondingly long acquisition times of up to 30 s per spectrum) were applied to avoid heating effects.

PLE mapping in air or vacuum   Fig. 5.1 presents typical PLE contour maps measured in air on top of several carbon nanotube forests. Rich PL peak patterns but relatively weak signals indicate that these maps stem from small ensembles of individual SWNTs. µ-PLE contour maps with 1-2 peaks, as shown in Fig. 5.2, were also frequently obtained. Observed emission linewidths of 9-14 meV are also characteristic for individual SWNTs suspended over prepatterned substrates and measured at room temperature [151, 153].

The excitation and emission positions of ~ 250 PL peaks were determined from about 50 maps measured in air on different surface sites of several VANTAs and plotted as filled squares in Fig. 5.3. Additionally, relatively weak features attributable to low- and

5 Spectroscopic Characterization of Individual Carbon Nanotubes

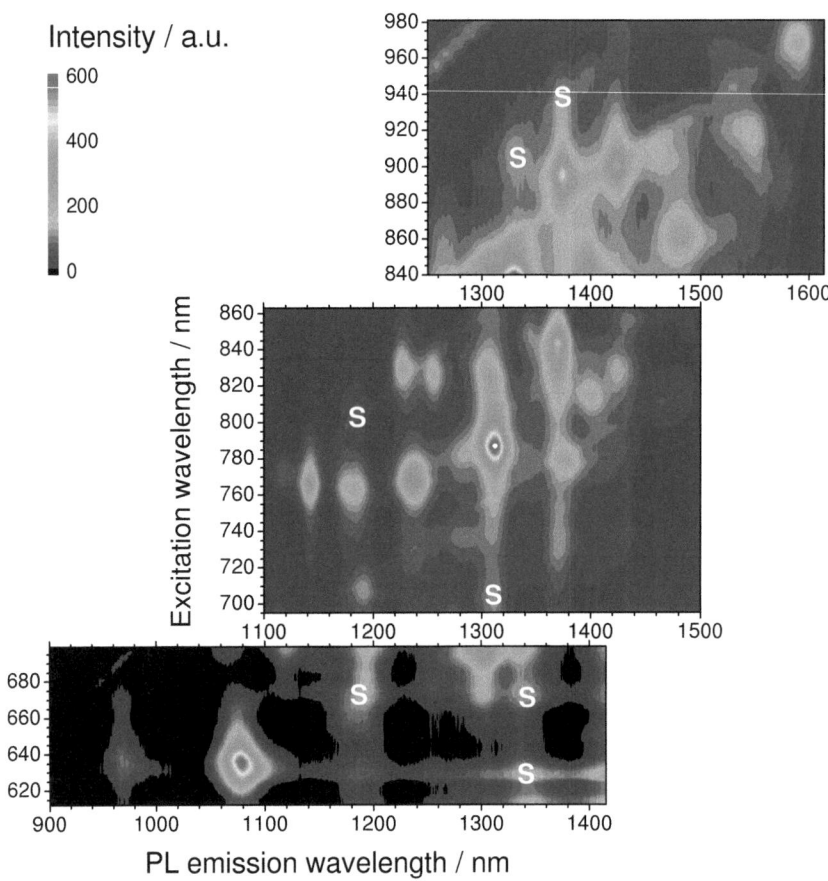

Figure 5.1: Typcial µ-PLE contour maps measured on top of a ∼ 0.2- mm-high carbon nanotube forest. The maps are measured at different sample sites within three overlapping excitation ranges. The ranges correspond to a dye and two Ti:sapphire lasers (see Sec. 3.1.2). Some PL features are assigned to low- and high-energy sidebands (S). Oblique stripes in the first and the last map are Raman bands of carbon nanotubes. From Ref. [9].

5.1 Influence of External Dielectric Screening on Optical Transition Energies

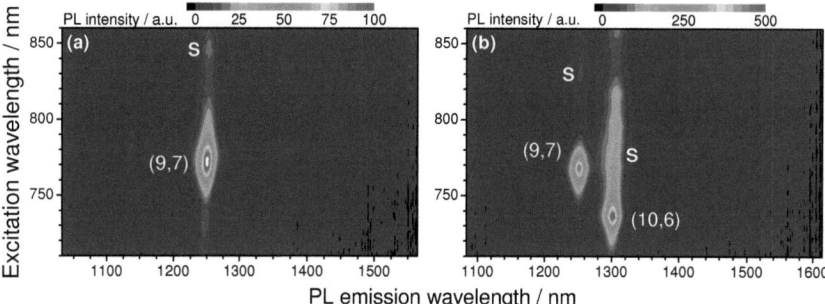

Figure 5.2: Typical µ-PLE contour maps measured on the top surface of a VANTA in air, demonstrating only few emitting SWNTs in the laser spot. The origin of the excitation sidebands (S) at 130-165 meV above the main peaks could be due to cross-polarized E_{12}/E_{21} absorption followed by E_{11} emission [35, 52]. We tentatively attribute the unusually shaped and pronounced sideband observed for the (10, 6) nanotube in **(b)** to interactions between this and the (9, 7) SWNT. From Ref. [9].

high-energy excitation sidebands were observed, marked with an (S) in Fig. 5.1 and 5.2. Some of these sidebands stem from exciton-phonon complexes, some others have not been assigned yet [52]. Such features were rejected in generating Fig. 5.3. Figure 5.3(a) also includes data points from one sample measured under a vacuum of $\sim 10^{-6}$ mbar. No systematic differences between PL energies measured in vacuum or under ambient conditions were observed. This result is due to the dielectric constant of air being close to unity. It also shows that gas adsorption onto the nanotubes at ambient conditions has only a negligible influence on the PL.

The open, red circles in Fig. 5.3(a) denote average PL peak positions for different (n_1, n_2) chiralities. An area enclosed by a circle corresponds to the scatter of experimental points, which was roughly $\pm(0.3\text{-}1)\%$ and $\pm(0.3\text{-}1.5)\%$ for E_{11} and E_{22}, respectively. This scatter is a lot larger than the measurement error and the uncertainty in assigning peak maxima, which we specify to less than ± 1 nm, corresponding to $\pm 0.1\%$ and $\pm 0.16\%$ for E_{11} and E_{22}, respectively. The larger spread of the PL data can be attributed to changes in the surroundings of emitting SWNTs. Following this assumption, we tentatively relate a few strongly scattered (redshifted) points in Fig. 5.3(a) to SWNTs detected slightly below the forest surface, in close proximity to other carbon structures.

77

5 Spectroscopic Characterization of Individual Carbon Nanotubes

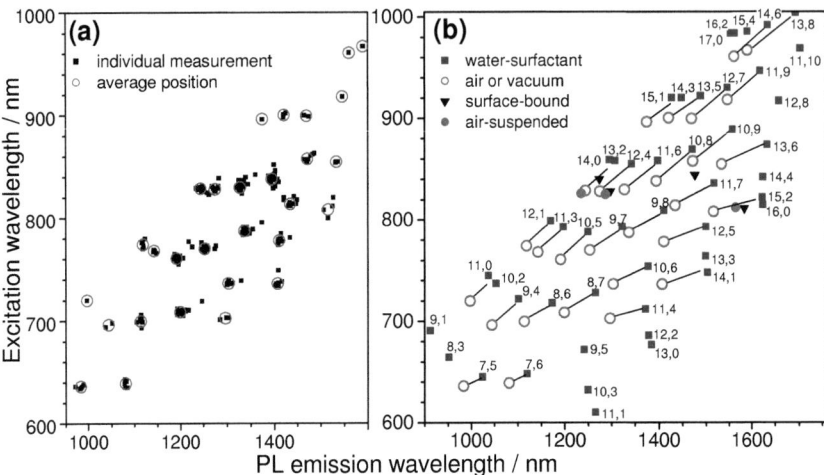

Figure 5.3: (a) Wavelength position of electronic transitions and average positions for ∼ 250 peaks, combined from μ-PLE maps of several samples of VANTAs measured in air and in vacuum. (b) Correlation of the PL peaks for SWNTs in air or vacuum to those in a water-surfactant dispersion and their assignment to specific (n_1, n_2) [100]. Filled red circles and black triangles correspond to measurements of individual, air-suspended and surface-bound SWNTs from Sec. 5.4.3 and 5.4.4, respectively. Adapted from Ref. [9].

The screening effect of these carbon structures may be the reason for the observed redshift of optical transitions of such SWNTs [38].

(n_1, n_2) assignment of PLE peaks in air or vacuum   Figure 5.3(b) compares the average PLE peak positions of SWNTs in air or vacuum with the positions of empirical $\lambda_{11}/\lambda_{22}$ fit values of Weisman and Bachilo [100] for SWNTs in aqueous SDS dispersions. These fit values accurately reproduce (within ±3 nm) the experimental PLE peak positions obtained for various SWNT materials dispersed with SDS and SDBS surfactants [2, 78, 100]. Figure 5.3(b) also demonstrates that the λ_{11} and λ_{22} wavelengths recorded in air or vacuum are systematically blueshifted relative to the aqueous dispersion. This correlation can be seen best for the PLE peaks of small-diameter nanotubes such as (7, 5) and (7, 6), which fall in a region of low (n_1, n_2) species density per wavelength or energy interval. Another 'anchor point' for the correlation are the small-helical-angle (zig-zag)

5.1 Influence of External Dielectric Screening on Optical Transition Energies

nanotubes which produce a 'diagonal edge' in the point distribution in Fig. 5.3(b). The blueshifts of PLE peaks amount to $+(40\text{-}56)$ meV and $+(24\text{-}48)$ meV with mean values of 50 ± 4 and 37 ± 7 meV for ΔE_{11} and ΔE_{22}, respectively. Table 5.1 lists the corresponding (average) excitation-emission wavelengths and energies as well as the observed shifts. Note that shifts are quantified only for those 19 (n_1, n_2) species for which at least three independent PLE measurements were obtained and for which the chiral index could be unambiguously assigned.

Within our experimental error or scatter, no systematic correlation between chiral index and variations in energy shifts was found. For instance, Fig. 5.4, representing ΔE_{11} and ΔE_{22} as a function of the nanotube diameter and the $(n_1 - n_2)$ mod 3 value, shows no such correlation. Surprisingly, ΔE_{11} which is determined more accurately than ΔE_{22}, is almost constant for all detected (n_1, n_2). We conclude that chirality dependent effects on the emission energy shifts are comparable to the uncertainty in determining ΔE_{11}, i.e. not larger than a few meV.

PLE mapping of immersed nanotubes The low-density VANTAs are mechanically quite delicate structures. For instance, they easily collapse irreversibly upon contact with tweezers. They also dramatically contract when rapidly immersed in liquids such as water, acetone or toluene due to capillarity forces [156]. In both cases, 'collapsed' material did not show any detectable PL. After storage under ambient conditions for several months, SWNTs on the surfaces of VANTAs were found to have become less luminescent, presumably for the same reason.

However, PL from individual SWNTs can still be detected after careful and slow immersion of the VANTAs into relatively viscous, nonpolar solvents such as paraffin oil and 1-methylnaphthalene ($\epsilon_r \approx 2-2,5$ in the GHz range, *vide infra*). Typical µ-PLE maps are shown in Fig. 5.5(a)-(c). PLE peak positions observed in 1-methylnaphthalene and especially in paraffin oil are close to those measured in aqueous dispersions, see Fig.5.5(d). The mean shifts for ΔE_{11} (ΔE_{22}) are -1 meV (-13 meV) and -14 meV (-27 meV) for paraffin oil and 1-methylnaphthalene, respectively.

This suggests a similar screening effect, i.e. a similar value of ϵ_r for the water-surfactant environment as compared to both organic solvents. We explain this in terms of the surfactant (i.e. hydrocarbon) coating around nanotubes as well as the dynamic character of the exciton creation and annihilation. At frequencies corresponding to the typical measured PL lifetimes of individual SWNTs ($\sim 20\text{-}200$ ps, corresponding to

Table 5.1: (n_1, n_2)-Assignment of optical transition energies E_{11} and E_{22} for SWNTs in air or vacuum and corresponding energy shifts ΔE_{11} and ΔE_{22} relative to SWNTs in water-surfactant (SDS, SDBS) dispersions. From Ref. [9].

(n_1, n_2)	λ_{11}, air (nm)[a]	λ_{22}, air (nm)	E_{11}, air (eV)	E_{22}, air (eV)	ΔE_{11} (meV)[b]	ΔE_{22} (meV)[b]	number of peaks measured	number of peaks rejected[c]
7,5	983(7)	636(2)	1.26	1.95	53	24	5	0
11,0/10,2[d]	997	720	1.24	1.72	-	-	1	0
9,4	1044	696	1.19	1.78	-	-	2	0
7,6	1081(2.5)	639(3)	1.15	1.94	44	24	5	0
8,6	1113(4)	700(4)	1.11	1.77	55	37	7	0
12,1	1118(2.5)	774(3.5)	1.11	1.60	51	34	7	0
11,3	1143(2.5)	768(1.5)	1.09	1.62	50	43	6	0
10,5	1191(6)	761(2.5)	1.04	1.63	49	48	20	2
8,7	1201(6)	709(2)	1.03	1.75	56	42	21	1
13,2/14,0[d]	1245(9)	828(3)	1.00	1.50	-	-	9	0
9,7	1253(4)	770(2)	0.99	1.61	52	37	25	2
12,4	1275(4.5)	828(2)	0.97	1.50	50	42	10	1
11,4	1296(9)	702(2)	0.96	1.77	54	26	3	0
10,6	1303(3.5)	737(1)	0.95	1.68	54	37	12	1
11,6	1329(6)	831(4)	0.93	1.49	48	39	20	1
9,8	1338(6)	787(2.5)	0.93	1.57	52	37	20	1
15,1/14,3[d]	1374	896	0.90	1.38	-	-	1	0
10,8	1395(6.5)	838(5)	0.89	1.48	48	44	20	0
13,3/14,1[d]	1408(5)	738(5)	0.88	1.68	-	-	7	0
12,5	1410(3.5)	778(2.5)	0.88	1.59	50	38	14	1
13,5	1420(3)	900(2)	0.87	1.38	40	31	3	0
11,7	1434(8)	814(3)	0.86	1.52	51	42	14	0
12,7	1468	900	0.84	1.38	-	-	2	0
10,9	1474(9)	858(3)	0.84	1.45	49	45	7	0
14,4/15,2/16,0[d]	1518(9)	810(8)	0.82	1.53	-	-	4	0
13,6	1533(3)	855(1)	0.81	1.45	50	35	3	0
11,9	1546	918	0.80	1.35	-	-	1	0
14,6(?)	1560	961	0.79	1.29	-	-	2	0

[a] average values and standard deviations (in brackets).

[b] energy shifts in air/vacuum vs. aqueous dispersion are indicated only if the shifts were determined from at least three measurements and unambiguously assigned to specific (n_1, n_2).

[c] redshifted PLE peaks, presumably due to screening (see text for details).

[d] PLE peaks can be assigned to a several (n_1, n_2).

5.1 Influence of External Dielectric Screening on Optical Transition Energies

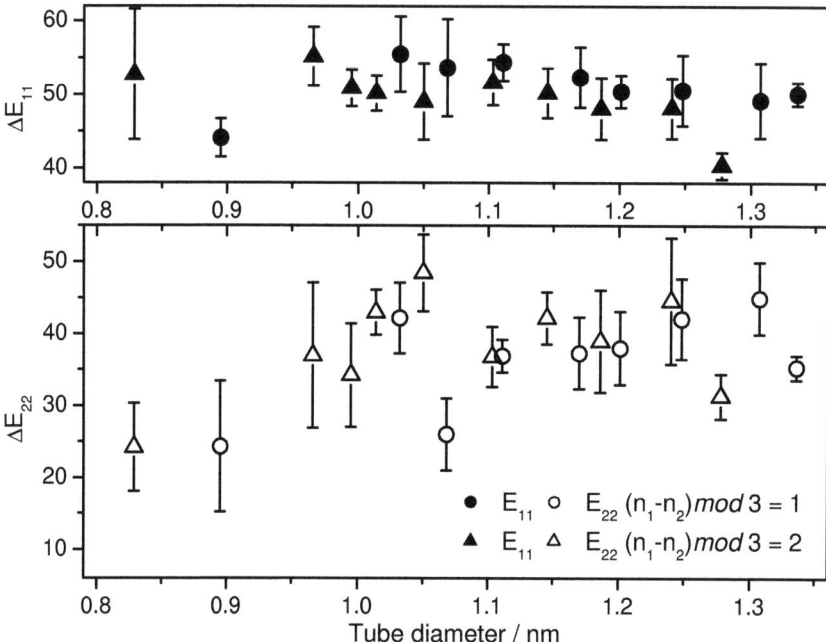

Figure 5.4: Shifts of E_{11} and E_{22} energies for semiconducting SWNTs in air/vacuum relative to aqueous dispersions (D$_2$O/ 1 wt% SDBS) as a function of nanotube diameter and $(n_1 - n_2)$ mod 3 value. Error bars show experimental uncertainty in determining the shifts. Experimental error of PLE peak positions in D$_2$O/SDBS is neglected. From Ref. [9].

50-5 GHz [59]) the real part of the permittivity of water is between 16 and 74 [157, 158] and decays further for faster relaxation processes. Additionally, with a diffusion length on the order of 100 nm [159, 160], the time a water molecule at a specific tube position is exposed to the electric field generated by an exciton is even lower than the PL lifetime. Thus, the effective frequency is higher than 50-5 GHz and ϵ_r of water becomes much smaller than its stationary value of $\epsilon_r \approx 81$ and closer to the permittivity of typical organic solvents at theses frequencies ($\epsilon_r \approx 2-3$).

The slight differences of $\Delta E_{ii}, i = 1, 2$ observed for SWNTs in different environments (including measurements on aqueous dispersions with different surfactants [78, 100] and

5 Spectroscopic Characterization of Individual Carbon Nanotubes

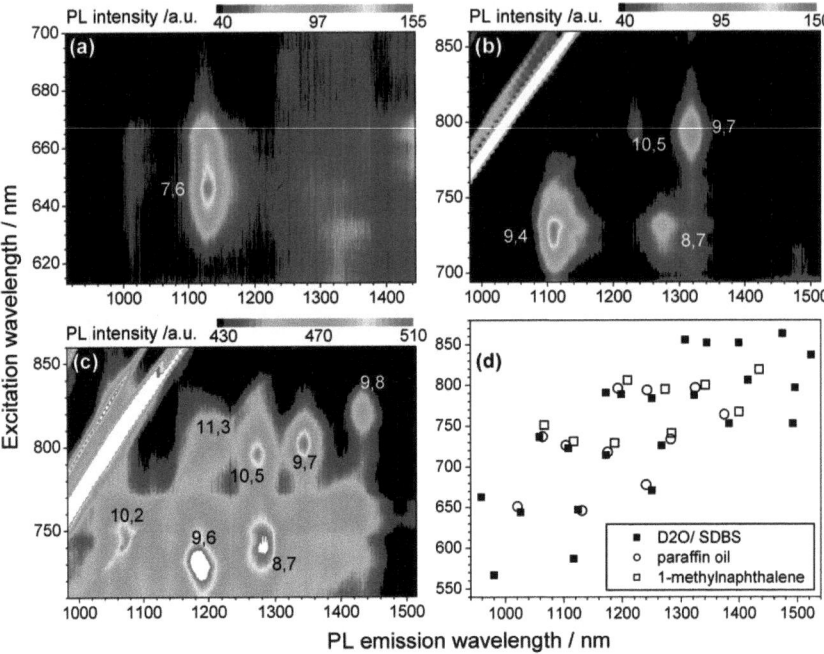

Figure 5.5: PLE of SWNTs immersed in organic liquids. **(a)** and **(b)** Exemplary μ-PLE maps of SWNTs immersed in paraffin oil and **(c)** in 1-methylnaphthalene. (n_1, n_2)-Assignment is based on their correlation with PLE peaks of SWNTs in a $D_2O/1$ wt% SDBS dispersion shown in **(d)**. Bright stripes are caused by Raman signals of carbon nanotubes and the organic liquids. **(d)** Comparison of PLE peak positions for nanotubes immersed in paraffin oil and 1-methylnaphthalene (averaged from several measurements) to HiPco SWNTs in $D_2O/1$ wt% SDBS. From Ref. [9].

our data for VANTAs immersed in decahydronaphthalene) are difficult to explain with a (uniform) external dielectric screening effect. They presumably indicate specific interactions of SWNTs with their surroundings. For example, electron transfer to SWNTs has been proposed for solvents containing strong electron-donating groups [161]. These interactions can also depend on the nanotube chirality.

Note that shifts in the PLE of immersed CVD-grown nanotubes as well as the absence of PL in compacted (collapsed) samples strongly suggests that only SWNTs and no multi-walled, e. g. double-walled nanotubes contribute to the PL. According to HRTEM and AFM analysis, most of the VANTAs are composed of large-diameter (≥ 2 nm) tubes with low-energy optical transitions, unaccessible in typical experiments. Still, we cannot exclude a minor fraction of MWNTs with small diameter inner tubes, structurally equivalent to emitting SWNTs. However, the inner tubes are 'shielded' and should therefore experience a quite different screening environment compared to SWNTs, and this should result in significant differences in their PL properties, if the inner tubes show PL at all. No such differences were found in this work.

5.1.3 Dielectric Screening Effect in SWNTs

In general, external dielectric screening seems to be the major reason for the PLE energy shift of SWNTs suspended in air or vacuum ($\epsilon_r \approx 1$) vs. SWNTs in water-surfactant or organic environments ($\epsilon_r \approx 2-3$). The observation of relatively *small* (a few tens of meV) and *positive* energy shifts in going from $\epsilon_r \approx 2-3$ to $\epsilon_r = 1$ is in good agreement with theoretical results of Ando [66] and other workers [64, 74, 150, 162]. These authors find that E_{11} and E_{22} optical transition energies correspond to the creation and annihilation of optically active excitons in the first and second van Hove singularities of semiconducting SWNTs. Therefore, the transition energies may be written as the difference of the single-particle (= band gap) energy E_{SP} and exciton binding energy E_b:

$$E_{ii}(\epsilon_r^{eff}) = E_{SP,ii}(\epsilon_r^{eff}) - E_{b,ii}(\epsilon_r^{eff}) \quad \text{with} \quad i = 1, 2 \tag{5.1}$$

with both terms depending on the effective screening constant ϵ_r^{eff}. The first exciton binding energy, $E_{b,11}$ derived from two-photon spectroscopy of SWNTs in an aqueous dispersion and in a polymer matrix ($\epsilon_r \approx 2-3$) is roughly inversely proportional to the nanotube diameter d and is ~ 0.27-0.42 eV for SWNTs with $d = 0.68$-1.18 nm [7, 8, 163]. The second exciton binding energy $E_{b,22}$ derived from resonant Raman spectra of electrochemically biased SWNTs is ~ 0.49 and 0.62 eV for SWNTs with diameters of 0.94 and 0.83 nm, respectively [164]. These experimental data correspond to calculated E_b energies for $\epsilon_r^{eff} = 2-4$ [74, 150], thus supporting the assumption that ϵ_r^{eff} is mostly contributed by the environment of the nanotube, i.e. $\epsilon_r^{eff} = \epsilon_r$. According to Perebeinos *et al.* [150], the exciton binding energy scales as $E_b \propto \epsilon_r^{-1.4}$ for $\epsilon_r \geq 4$. When extrapolating this relation to $\epsilon_r = 1$ for SWNTs in air, exciton binding energies

$E_{b,11} \approx 0.8\text{-}1\,\text{eV}$ are obtained. However, the single-particle energy is simultaneously enhanced (also called 'band gap renormalization') and overcompensates this significant increase in E_b [66, 162]. Both result in a relatively small increase of the E_{11} and E_{22} energies as observed in our experiments.

A realistic limit for external dielectric screening of SWNTs may be $\epsilon_r = 5$. The energy shifts of SWNTs embedded in such (inorganic) media vs. aqueous dispersion are expected to be small as well—comparable to those of SWNTs in air but opposite in sign. However, the effect of increased screening on the exciton relaxation rate and PL efficiency could be very pronounced. The low PL efficiency ($\sim 10^{-3} - 10^{-2}$) of SWNTs has been explained by fast relaxation to the dark exciton state(s) $\sim 90\,\text{meV}$ below the optically active state responsible for the PL [74]. Assuming a similar scaling relation as for E_b, this energetic difference would be only $\sim 25\,\text{meV}$ at $\epsilon_r = 5$, i.e. comparable with the thermal energy at room temperature. Consequently, a thermal repopulation of the optically active excitonic state could be expected, resulting in strongly enhanced PL.

Exciton diffusion to low-energy areas So far, our interpretation implicitly assumes that excitons not only arise, but also decay in air-suspended sections of SWNTs. However, another possibility might be the diffusion of relatively long-lived E_{11} excitons along a nanotube to domains with lower exciton formation energies. If the associated gain in energy overcomes thermal energy, excitons would be trapped there. Such domains might correspond to nanotube sections (e.g. ends) which are screened by the substrate or other supporting structures. This local screening-induced diffusion of excitons in SWNTs would be analogous to strain gradient-induced diffusion of excitons in pure silicon crystals at low temperatures as studied by Tamor and Wolfe [165]. As opposed to bulk silicon, expected exciton formation energy gradients in SWNTs can be on the order of tens of meV, i.e. comparable to thermal energy at room temperature.

Exciton diffusion to low-energy, screened domains followed by recombination and PL emission would significantly complicate the determination of energy shifts corresponding to $\epsilon_r = 1$. However, we exclude this effect in our samples based on the following arguments. Most PLE peaks recorded on different sample sites of several samples and assigned to the same (n_1, n_2) chirality are relatively close to each other [Fig. 5.3(a)]. This indicates a similar surrounding for recombining excitons whereas the opposite is expected for trapped excitons. Moreover, diffusion and trapping of excitons in different domains of the same suspended SWNT (each suspended nanotube has at least two contacts with the support) would statistically produce PL signals with slightly varying emission energies which we did not observe. On the contrary, PLE maps with a single, narrow emission peak were often measured [see, e.g. Fig. 5.2(a)].

The conflicting results obtained so far by different research groups mentioned in the beginning suggests a need for further experimental work on environmental effects in SWNTs. The small blueshifts of E_{11} and E_{22} energies found by Lefebvre et al. [151] for individual SWNTs grown over submicrometer high structures in silicon might be due to screening effects from the substrate. In this context, SWNTs grown by Yin et al. on substrates with 1-2 µm wide, deep trenches seem to be more suitable. However, their recent Raman investigation of similar samples suggested that the majority of nanotubes were likely grown in small bundles and that intertube screening is responsible for the large redshifts of E_{22} energies [166]. E_{11} and E_{22} energies reported by Ohno et al. for SWNTs grown over a grooved substrate are on average blueshifted, but they contain a component whose sign and magnitude depends on the value of $(n_1 - n_2)$ mod 3 and the nanotube chiral angle θ [153]. This component is very similar to E_{11}/E_{22} shifts expected for axially stretched nanotubes [167], suggesting that the nanotubes experienced strain by a thermally expanding or contracting substrate. This could have been caused e.g. by laser heating.

5.2 Blinking of SWNTs at Cryogenic Temperatures

5.2.1 Blinking Background

In single molecule spectroscopy, phenomena like temporal fluctuations of PL intensity, so-called PL intermittency or simply 'blinking', as well as spectral fluctuations (= shifts in emission peak position), also termed 'spectral diffusion' or 'spectral wandering', are quite common [168]. Both phenomena have been observed at low temperatures for individual organic molecules ([169], and references therein), single polymer strands [170], semiconductor quantum dots [171, 172] and noble metal clusters [173]. The observation of blinking and spectral diffusion is normally regarded as an indication that single nanoscale objects are being probed, as opposed to ensembles.

The processes of PL blinking and spectral diffusion depend on the system under investigation and are in general not completely understood. The blinking in semiconductor quantum dots is believed to be due to trapping of photoexcited electrons or -holes on the surface or within the matrix surrounding the quantum dot. In the so-ionized quantum dot, nonradiative Auger processes quench all luminescence [174, 175] until the trapped charge returns to the quantum dot. The mechanisms of PL blinking and spectral wandering in SWNTs are also still part of ongoing research.

Htoon *et al.* observed both types of single molecule phenomena for the PL of individual SWNTs deposited onto quartz from a water-surfactant dispersion at a temperature of 4 K [176]. Their nanotubes show PL blinking including complete 'on-off' switching and spectral diffusion of up to ~ 20 meV on a time scale of seconds. Note that individual SWNTs prepared from water-surfactant dispersions are still coated with surfactant molecules to a certain degree, even though the sample is usually rinsed with the solvent. This was shown by AFM image analysis [177]. Rigorous rinsing causes the nanotubes to be removed from the substrate. SWNTs prepared in a similar way by Hartschuh *et al.* did not show any PL fluctuations at ambient temperature [178] whereas Matsuda *et al.* did observe blinking for some tubes even at room temperature [179]. This discrepancy could either be attributed to very fast PL fluctuations at ambient temperatures or to chemical/structural defects in the surfactant/the SWNT itself acting as charge traps [179]. The only PL study of CVD-grown SWNTs at low temperatures stems from Lefebvre *et al.* They grew air-suspended, individual SWNTs (i.e. free of surfactant) on prepatterned silicon substrates. Surprisingly, they observed complex PLE contour maps with a lot of unusual temperature-dependent PL features that remained unassigned [180]. In this section and in Ref. [10], we compare the PL of SWNTs deposited from a water-surfactant dispersion to CVD-grown nanotubes from room temperature down to 4 K. We find PL intermittency and spectral diffusion only for the former at low temperatures. CVD-grown SWNTs show stable emission over minute time scales at all temperatures. We therefore believe that PL intermittency and spectral diffusion in SWNTs are not intrinsic, but can be caused by interactions with a surfactant coating. In addition, we demonstrate that PLE spectra of well-separated SWNTs at low temperatures are much simpler and manifest phonon-assisted exciton sidebands with lower intensity than spatially contacting nanotubes. This suggests that interactions between contacting SWNTs may have contributed to the complexity of low-temperature PLE spectra reported by Lefebvre *et al.* and Htoon *et al.* [84, 180] [1].

5.2.2 Sample Preparation and Measurements

The following samples of individual SWNTs were used for this study:

- Vertically aligned carbon nanotube arrays (VANTAs) comprising of MWNTs as supporting structures and a small fraction of SWNTs suspended between them (Sec. 4.1)

- Short pieces of SWNTs protruding from the catalyst pad of horizontally aligned SWNTs, some presumably in direct contact with a sapphire substrate and some air-suspended between cracks of the catalyst, grown by carbon monoxide CVD (Sec. 4.2).

- SWNTs from the HiPco-process, spin-coated on a sapphire substrate from an aqueous dispersion containing 1 wt% sodium cholate (Sec. 2.5.3).

Sapphire was the preferred substrate because it does not show any background PL as opposed to silicon wafers [e.g. Fig. 4.10(d)]. Samples were placed in a continuous-flow liquid helium microscope cryostat (CryoVac) and evacuated to $\sim 10^{-7}$ mbar. The cryostat was mounted underneath the microscope on a (x, y)-translation stage and a

---

[1] In Sec. 5.4.4, we show PL intermittency at room temperature for a CVD-grown SWNT after severe defects induced by AFM (tube rupture and bending). This demonstrates that defects can play a role, as well.

NIR objective (40x/0.6) was used for focusing. Spectral resolution depended on the diffraction grating used and was either ~ 1.5 or $0.3\,\text{meV}$. As mentioned in the preceding section, heat dissipation of suspended SWNTs in vacuum can become a problem at high power densities. A redshift of PLE signals can be observed if the effective nanotube temperature increases [155]. To avoid such heating effects, low laser powers and power densities of $\sim 0.1\,\text{mW}$ and less than $10\,\text{kW/cm}^2$, respectively, were used for suspended SWNTs. A reduction of laser power by another order of magnitude did not reveal any changes in the PLE contour maps, except for a decrease of signal-to-noise ratio. Due to a better heat dissipation, excitation powers could be increased to $1\,\text{mW}$ for individual, CVD-grown SWNTs on sapphire and $2\text{-}3\,\text{mW}$ for surfactant-coated SWNTs spin-coated on sapphire. Typical PL integration times were $10\,\text{s}$ per spectrum and 15-30 minutes for a complete PLE contour map.

5.2.3 Observations and Conclusions

Surfactant-coated SWNTs Typical low-temperature PLE maps (here in the form of 3D plots) of individual, sodium cholate-coated HiPco SWNTs deposited on sapphire are depicted in Fig. 5.6(a) and (b). The irregular, 'jagged' PLE profile, in particular for the (9,8) SWNT in Fig. 5.6(a), can be attributed to blinking during the measurement. Fig. 5.6(c) and (d) show 'time traces' of the same SWNTs at constant excitation wavelengths [805 nm and 786 nm for (9,8) and (9,7), respectively], where sequential PL emission spectra are combined to contour maps of time vs. emission wavelength. Emission energies vary by $\Delta E_{11} = 1.5$ and $0.5\,\text{meV}$ for (9,8) and (9,7), respectively. Similar fluctuations of the PL emission energy were seen below $\sim 25\,\text{K}$ for all ~ 70 individual SWNTs measured in spin-coated samples. Most SWNTs demonstrated spectral shifts close to (9,8), with ΔE_{11} up to $2\,\text{meV}$. Only a few nanotubes showed less pronounced (and likely faster) PL fluctuations, similar to the (9,7) SWNT. Apart from individual tubes, we also detected temporal and spectral fluctuations of small ensembles within the resolution of the objective ($\sim 0.7\,\mu\text{m}$, see Tab. 3.1). All PL emission peaks were symmetric with fwhm of 1-2 meV, virtually independent of temperature in the range of 4-10 K.

The temporal and spectral fluctuations mentioned above correspond qualitatively quite well to those reported by Htoon et al. [176]. They deposited HiPco SWNTs from an aqueous SDS dispersion on crystalline quartz and recorded PL spectra at $4\,\text{K}$. There

5.2 Blinking of SWNTs at Cryogenic Temperatures

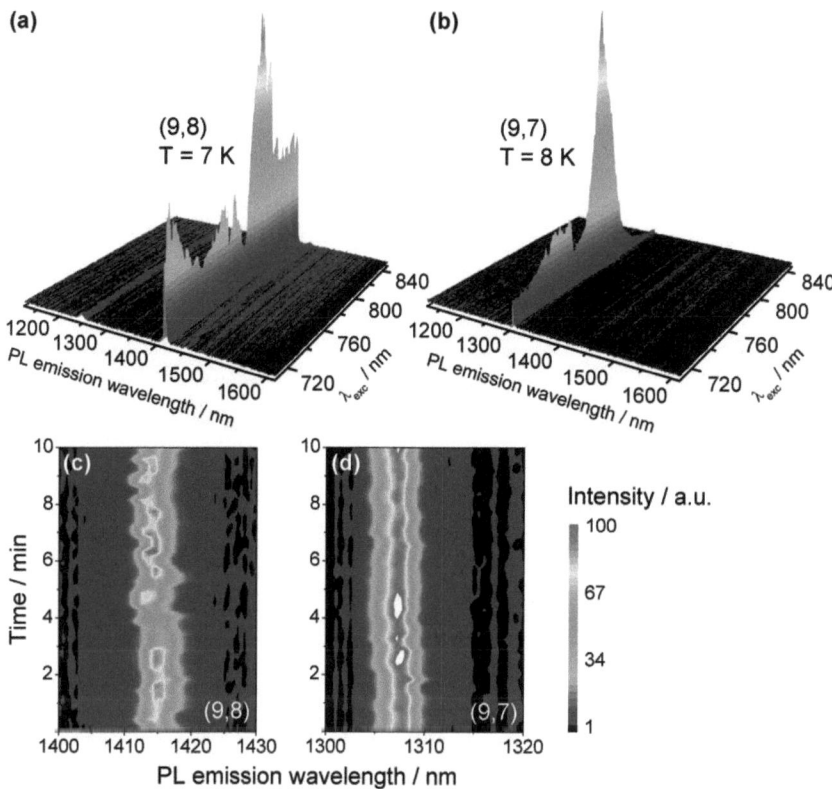

Figure 5.6: PL of SWNTs deposited on sapphire from an aqueous sodium cholate dispersion at 7 and 8 K. **(a) and (b)** 3D µ-PLE map of a (9, 8) and a (9, 7) SWNT. There are major—presumably phonon-assisted—sidebands in the vicinity of the electronic transition, represented by the strongest peak. **(c) and (d)** 60 sequential PL emission spectra, each with an acquisition time of 10 s at an excitation wavelength of 805 and 786 nm for (9, 8) and (9, 7) nanotubes, respectively. Assignment follows Ref. [100]. From Ref. [10].

is particularly one kind of SWNTs in their report which demonstrated blinking and spectral diffusion of a few meV as well as symmetric emission lines with fwhm below 5 meV. Quite contrary to this are nanotubes which exhibit jumps in ΔE_{11} of up to 20 meV, strongly asymmetric PL emission line shapes and fwhm of up to 10 meV. These features were attributed to accidental doping of SWNTs via gas adsorption and defects induced by extensive sonication [176]. We dispute this explanation based on the following argumentation. The SWNT material (HiPco) and the sonication procedure applied in this work were similar to those reported by Ref. [176]. However, we did not observe broad emission peaks or asymmetric lineshapes. Also, measurements on individual CVD-grown SWNTs under ambient conditions (mentioned in the previous section and in Ref. [9]) and at low temperatures (this section and in Ref. [10]) demonstrate that the influence of gas adsorption on the PL of SWNTs is negligible.

A different characteristic feature of deposited, surfactant-coated SWNTs which is also typical for PL measurements on ensembles of aqueous SWNT dispersions is the occurrence of strong PLE sidebands, i.e. signals below or above the main electronic E_{22} resonance transition [52, 83]. We observed such sidebands for deposited, sodium cholate-coated SWNTs at ambient as well as at low temperatures, in accordance with Htoon et al. who used SDS-coated nanotubes [84]. Some of these features have been attributed to exciton-phonon complexes, others have not been assigned yet. As mentioned in Sec. 2.4.4, such phonon-assisted PLE sidebands are believed to be a signature of strong exciton-phonon coupling in SWNTs [52, 82–84].

Surfactant free SWNTs In contrast to surfactant-coated SWNTs, constant PL emission (within the stability of our µ-PLE set-up) was observed for all CVD-grown, surfactant-free SWNTs at temperatures between ambient and 4 K, irrespective of the time scale (varied from seconds to minutes) and irrespective whether individual emitters or small ensembles of SWNTs were probed. We recorded about 50 µ-PLE spectra from two VANTA samples and one grown by carbon monoxide CVD. Two typical low-temperature 3D PLE spectra as well as corresponding time traces are shown in Fig. 5.7. They demonstrate stable PL emission and no spectral wandering for a $(8,7)$ SWNT measured on top of a VANTA in Fig. 5.7(a) and for a $(9,4)$ nanotube grown on sapphire by carbon monoxide CVD in (b).

Apart from stable PLE spectra, the SWNTs in Fig. 5.7 possess unusually weak PLE sidebands as compared to surfactant-coated nanotubes. We frequently found such simple 'one-peak' spectra for well-separated SWNTs. From 50 PLE spectra acquired from CVD-grown nanotubes, about 25 showed only one major signal corresponding to one chirality present in the detection area. Out of these 25, 20 nanotubes or 80% had weak PLE

5.2 Blinking of SWNTs at Cryogenic Temperatures

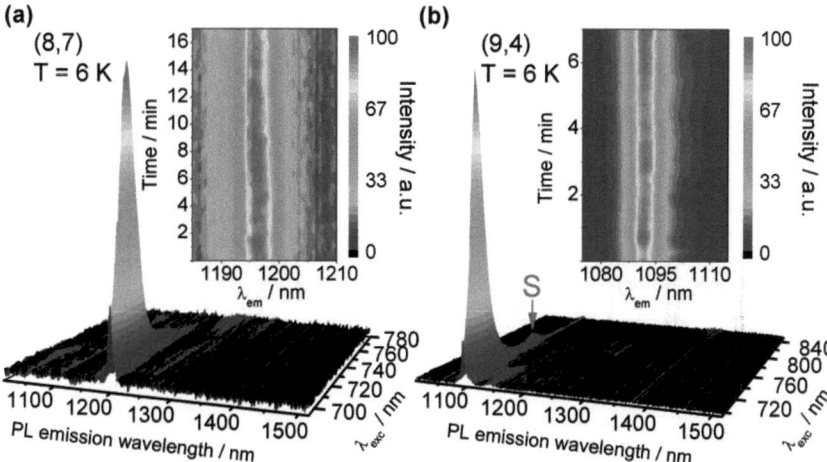

Figure 5.7: PL of individual, surfactant-free SWNTs grown by CVD at 6 K. (a) 3D μ-PLE map of a (8,7) SWNT suspended on top of a VANTA. The inset shows a time trace combined from 100 sequential spectra each acquired during 10 s at an excitation wavelength of 705 nm. (b) (9,4) SWNT grown on sapphire by carbon monoxide CVD. (S) denotes a weak PLE sideband. The inset is again a time trace combined from 40 sequential spectra each acquired during 10 s at an excitation wavelength of 710 nm. From Ref. [10].

sidebands. Two further examples from this fraction are depicted in Fig. 5.8.

Observed fwhm were independent of temperature in the range from 4 to 10 K and rather uniform with ∼5 meV in emission and ∼30 meV in excitation for air-suspended SWNTs. Seven μ-PLE spectra recorded from individual emitters were likely to stem from SWNTs lying on the sapphire substrate, close to the catalyst pad. This was confirmed by SEM and optical images, a lowered PL intensity (see also Sec. 5.4.4), additional shifts of E_{11} and E_{22} due to dielectric screening of the substrate compared to air-suspended nanotubes and by slightly broader emission linewidths of ∼7 meV. The (12,4) SWNT shown in Fig. 5.8(b) is an example of such a tube. In general, fwhm of PL emission signals at low temperatures are less than those measured at ambient temperature (∼9-14 meV, see previous section). However, they are still greater than thermal energy (∼0.3 meV at 4 K). This might be due to slightly different environments experienced by excitons which can diffuse a certain distance (on the order of 100 nm [159, 160]) before recombining.

5 Spectroscopic Characterization of Individual Carbon Nanotubes

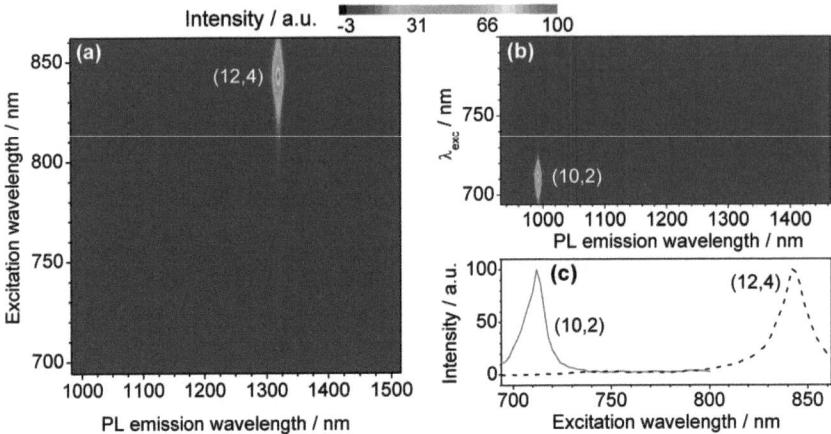

Figure 5.8: Further examples of PL at low temperatures from individual, surfactant-free SWNTs grown by CVD. **(a)** Contour map of a (10, 2) SWNT at 6 K suspended on top of a VANTA. **(b)** (12, 4) SWNT grown by carbon monoxide CVD in contact with the sapphire substrate at $T = 4.8$ K. **(c)** PLE profiles (i.e. vertical cross sections of a contour map) at the emission wavelengths of 992 and 1320 nm for the (10, 2) and the (12, 4) nanotube shown in **(a)** and **(b)**, respectively. Sidebands to the optical transitions are hardly observed. From Ref. [10].

A comparison of the PLE data for surfactant-coated and surfactant-free SWNTs suggests that blinking and spectral wandering are of 'extrinsic' origin and related to a particular SWNT environment (or defects). However, the exact mechanism is not clear yet. In analogy to related systems exhibiting PL blinking (see Sec. 5.2.1) we propose that photoexcited charges from the SWNT are trapped in the surfactant layer and slowly recombine on a time scale of seconds at a few Kelvin. Such external charges can cause spectral wandering due to a Stark shift of the emission energy [169] [2]. Blinking has so far been observed for SWNTs coated with SDS [176, 179] and sodium cholate surfactants [182]. The chemical structure of both compounds is quite different except that they are ionic sodium salts. One might expect similar effects for SWNTs enclosed in different organic matrices. Whether the ionic character of these surfactants is important or not is unknown. However, a purely neutral encapsulation by atmospheric gas molecules

---

[2] In this context, it is interesting to note that the PL of nanotubes can be quenched by applying external electrical fields [181]

adsorbed onto cold CVD-grown nanotubes does not lead to spectral or temporal PL fluctuations.

A similar concept of ejected charges trapped in a dielectric environment has been used to explain hysteresis effects in the gate-modulated conductance of semiconducting SWNTs, as well [183]. The hysteresis complicates the use of SWNTs in future electronic devices. If fluctuations in conductance and PL have the same origin, PL spectroscopy could be used to check the quality of SWNT-based transistors.

The stable, low-temperature PL of surfactant-free SWNTs demonstrated in this subsection are quite consistent with observations of Lefebvre et al. for CVD-grown SWNTs under helium, bridging submicrometer-high structures on silicon at temperatures down to 5 K [180]. However, in contrast to this work, they observed rather complex spectra with peak splittings of a few meV and small shifts of emission lines upon decreasing the temperature below 20 K. One possible explanation for these effects could be a partial localization and radiative recombination of excitons in regions of the nanotube screened by the substrate or other structures with lower exciton formation energies (we discussed this possibility in the previous section). This could explain the onset of splittings of a few meV at about 20 K. Despite this difference which is likely caused by the different kind of samples, the observation discussed in the next paragraph is also in accordance with the results of Lefebvre et al.

Small ensembles of SWNTs When several individual emitters were within the detection area of our microscope, we frequently observed much more complicated structures than just a superposition of individual PLE signals. Typical features included (i) broad and strong PLE sidebands, (ii) enhanced phonon-assisted PLE sidebands and (iii) unusually shifted and unidentified PLE peaks. Due to the smaller linewidths at cryogenic temperatures, these features were particularly apparent. Figure 5.9 shows an example of such a case for nanotubes detected on top of a carbon nanotube forest at 6 K. Three air-suspended, individual SWNTs could be identified and assigned to (13, 2), (12, 4) and (8, 6). There are high-energy PLE sidebands (S) ~ 100 meV and 180 meV above the zero-phonon line (white ovals) with signal intensities up to 20%. The latter could be due to a exciton-phonon complex of a dark exciton, ~ 20 meV below the lowest bright E_{22} exciton and a G-mode phonon [84, 85]. The sideband assigned as $E_{11} + (D^* + G)$ of the (12, 4) SWNT is even brighter than the zero-phonon line, although the D^* and $(D^* + G)$ Raman signals are rather weak. In addition, there are several features in

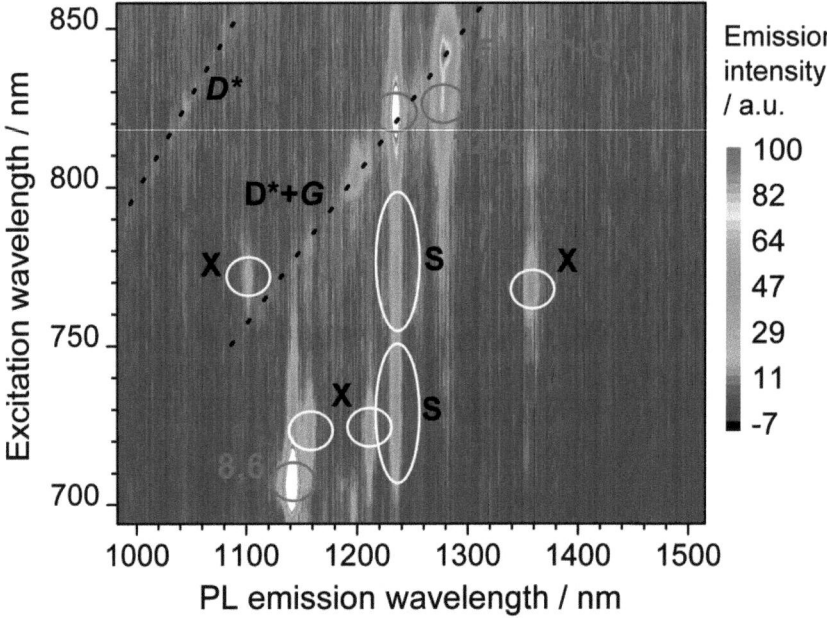

Figure 5.9: µ-PLE contour map from a small ensemble of individual, closely adjacent surfactant-free SWNTs grown by CVD and detected on top of a VANTA at $T = 6\,\text{K}$. Red open circles are assigned to pure electronic transitions of air-suspended $(13, 2)$, $(12, 4)$ and $(8, 6)$ SWNTs. Dotted lines correspond to D^* and $(D^* + G)$ Raman bands. Note that the $E_{11} + (D^* + G)$ signal of the $(12, 4)$ nanotube is stronger than that originating from the zero-phonon exciton. White ovals (S) denote high-energy $(E_{22} + \Delta E)$ PLE sidebands of the $(13, 2)$ SWNT. Unidentified PL features are indicated by white open circles (X). From Ref. [10].

Fig. 5.9 marked with open white circles which could not be unambiguously assigned. Their intensity is comparable to PLE sidebands. If they originated from SWNTs with other chiral indices, their transition energies would be unusually shifted compared to suspended, substrate-screened or dispersed SWNTs (*vide infra*).

Our results for individualized CVD-grown SWNTs as compared to small ensembles thereof suggest that spatially close-lying segments of individual SWNTs in ensembles have a substantial influence on each others PLE spectra, like the enhancement of PLE sidebands. Therefore, one can conclude that the strength of exciton-phonon coupling depends on the environment of the SWNT and the thereby introduced perturbations. One kind of perturbation can be constituted by intertube interactions. In contrast to the special case of Fig. 5.9, exciton-phonon coupling appears to be especially weak in air-suspended and therefore unperturbed SWNTs at low temperatures, as demonstrated by Fig. 5.7 and 5.8.

(n_1, n_2)-assignments and PL intensities The assignment of E_{11}/E_{22} optical transition energies to specific (n_1, n_2) chiralities given throughout the section is based on different references and observations: At ambient temperature, PLE maxima of deposited surfactant-coated SWNTs are close to the ensemble values of SWNTs in the initial dispersion, albeit with some scattering as individual emitters are probed (this topic is discussed in Ref. [111]). Thus, PLE signals can be assigned to (n_1, n_2) chiralities according to Weisman and Bachilo [100]. Referenced to a dispersion of SWNTs, air-suspended and surfactant-free nanotubes demonstrate a relatively uniform increase of $\Delta E_{11} = 50\pm4$ meV and $\Delta E_{22} = 37\pm7$ meV due to the absence of dielectric screening, as discussed in the previous section [9]. These shifts and the assignment of Weisman and Bachilo can also be applied at cryogenic temperatures if taking into account a small additional blueshift of a few meV as a result of the temperature decrease [155, 180]. However, for tubes with very similar E_{11}/E_{22} energies, an unambiguous assignment was not always possible. The external dielectric screening effect of bare SWNTs on sapphire substrates is smaller than that for dispersed or deposited, surfactant-coated SWNTs, but obviously not as small as air-suspended tubes with $\epsilon_r = 1$. Consequently, the shifts are in between both, and referenced to dispersed SWNTs, there is a blueshift of $\Delta E_{11} \approx 15 - 30$ meV and $\Delta E_{22} \approx 10 - 25$ meV. The matter of surface-bound SWNTs (on SiO_2) and the shifts involved will be discussed further in Sec. 5.4.4.

For a comparison of PL intensities of SWNTs in different environments (air-suspended, deposited from dispersion and in direct contact with a surface) at ambient temperatures, the maximum intensity from a PLE contour map of an individual SWNT, measured with the same microscope set-up, was divided by acquisition time and excitation power. We thereby assume (and checked) that the relation of PL intensity and excitation power is still in a linear range. Other calculation schemes like using the integral PL intensity (an area) when the nanotube is excited at E_{22} resonance or even the intensity integrated over the whole 2D PLE map (a volume) with or without fitting the peak with analytical 1D

5 Spectroscopic Characterization of Individual Carbon Nanotubes

or 2D functions were tested but resulted in roughly the same results. It turns out that the individual scatter from one measurement/SWNT to the other (even with the same chiral index) is larger than the error of the different calculation schemes. We see this as a further evidence that the environment (different configurations, defects, external perturbations etc.) of a nanotube has paramount influence on the PL intensity and therefore on its PL quantum yield. Of course, other factors like the orientation of the polarization vector relative to the tube axis are important, as well. On average, typical PL intensities were found to scale as $\sim 100 : 50 : 5$ for individual, air-suspended SWNTs; surfactant-coated, deposited SWNTs and as-grown, surface-bound CVD SWNTs (on sapphire or SiO_2). This relation holds also at cryogenic temperatures. At low temperatures, the integral intensities increased by a factor of $\sim 2-5$ whereas the absolute intensity on the emission/absorption peak maximum increases by about one order of magnitude which is due to narrower PL peaks.

In Sec. 5.5 we show measurements of single nanotubes randomly diffusing in a water-surfactant dispersion [13]. Their intensities are comparable to air-suspended SWNTs, but we used a 100x/1.4 oil-immersion objective (Leica) instead of a non-contact 100x/0.95 objective (Olympus), yielding a higher excitation power density in the focus and a larger collection efficiency (a numerical aperture of 1.4 as compared to 0.95). This accounts approximately for a factor of two ($1.4^2/0.95^2$) and shows that SWNTs deposited from a water-surfactant dispersion possess similar PL intensities as in dispersion.

5.3 Direct Observation of Deep Excitonic States in the PLE Spectra of SWNTs

5.3.1 Bright and Dark Excitonic Levels

It has been mentioned in Sec. 2.4.4 that for the excitonic picture of SWNTs, interband, single-particle transitions are conceptually replaced by manifolds of excitonic states progressing to a corresponding continuum band [see also Fig.2.8(c) and (d)]. Absorption (more specific: one-photon absorption, OPA) and PL emission is then explained on the basis of exciton creation and annihilation into/from dipole allowed ('bright') states. Additionally, group theory and theoretical calculations predict dipole forbidden, so-called 'dark' states below every bright exciton, including the lowest one. However, even the most recent calculations cannot unambiguously determine the energy separations of exciton manifolds, relaxation dynamics and the chirality dependence. For instance, calculated separations of the lowest (singlet) dark excitonic level below the lowest bright E_{11} exciton range from about -5 meV to more than -100 meV [8, 56, 73–77]. Note that in our case, 'bright' and 'dark' states refer to one-photon absorption (OPA) and emission only. The bright (dark) excitonic states corresponding to the interband transitions formally labeled as E_{ii} are now labeled according to wave vector group theory as A_2 (A_1) for chiral and A_{2u} (A_{1u}) for zig-zag SWNTs, respectively (armchair nanotubes are always metallic and not part of this work). However, these labels do not permit a differentiation between the bright exciton of the E_{11} and E_{22} band. Therefore, we call the $A_{2(u)}$ exciton of band i 'optically allowed' E_{ii} exciton or OE_i for short. Weakly optically allowed excitons are referred to as 'dark' or 'weakly allowed' excitons or DE_n, where n is a running index of the different dark excitonic states observed. As this work is purely experimental, we do not consider a further assignment of the DE_n states to $A_{1(u)}$ or to e.g. triplet excitons which are also below OE_i [77].

So far, the existence of deep excitonic states in SWNTs has not been observed directly. Such states have been postulated to explain the low PL quantum yield [8, 73, 74] ($\eta \approx 10^{-3}$ from ensemble measurements [2, 78]) and the observations from pump-probe experiments that specific absorption transients decay more slowly than the PL [184, 185]. For the interpretation of the PL temperature dependence of SWNT dispersions, a shallow dark state only 1-5 meV below OE_1 has been invoked [186, 187].

5 Spectroscopic Characterization of Individual Carbon Nanotubes

Here and in Ref. [11], we report for the first time the observation of red-shifted PL emission satellites of individual, air-suspended SWNTs which we tentatively assign to weakly optically allowed excitonic states approximately 38-45 and 100-130 meV below OE_1. The nanotubes were excited via OE_2 using µ-PLE spectroscopy. The same experimental set-up, cryostat and objective as already described in Sec. 5.1 and 5.2 was used. Similar satellites redshifted by 90-143 meV for different (n_1, n_2) species were also found for ensembles of dispersed SWNTs which were excited via OE_1. Excitation of ensembles via OE_1 has the advantage that the satellites do not overlap with features from other chiralities.

The samples used in this study were similar to those in the previous section: (i) CVD-grown VANTAs and in particular the SWNTs air-suspended on the top surface; (ii) SWNTs grown by carbon monoxide CVD, air-suspended between cracks in the catalyst material and (iii) HiPco and PLV nanotubes in D_2O/SDS or SDBS dispersions. PLV-SWNT dispersions with a very narrow chirality distribution were obtained by density gradient ultracentrifugation in iodixanol [33]. Ensemble PL spectra of dispersions were acquired with a Fourier Transform IR spectrometer (Bruker, 'IFS66') equipped with a liquid-nitrogen-cooled Ge photodiode [78]. For excitation, a monochromatized xenon lamp light source with a fwhm bandwidth of 7.5 nm and a wavelength scan step of 3 nm was used. FTIR-PL maps were corrected for wavelength-dependent excitation intensity and spectrometer response function.

5.3.2 Observations and Conclusions

In order to facilitate interpretation of the PLE spectra, only single emitting SWNTs, preferably also ones having a high emission intensity were studied on every sample. We believe that most SWNTs satisfying this criteria were air-suspended (see previous section or Ref. [10]). Figure 5.10(a) shows an example of such a single $(9,7)$ SWNT at ambient conditions. A relatively intense, single peak PLE spectrum and a characteristic blueshift compared to a $(9,7)$ tube in a water-surfactant dispersion indicates an individual, air-suspended SWNT (see Sec. 5.1 or Ref. [9]). Apart from a sideband (S) at the same emission energy, there are also two weak and broad emission satellites redshifted by ~ 40 meV (shoulder) and ~ 110 meV. We focus on these emission satellites in this study and label the former DE_1 and the latter DE_2. Similar features have been observed in four other single-emitter PLE contour maps which exhibited a

5.3 Direct Observation of Deep Excitonic States in the PLE Spectra of SWNTs

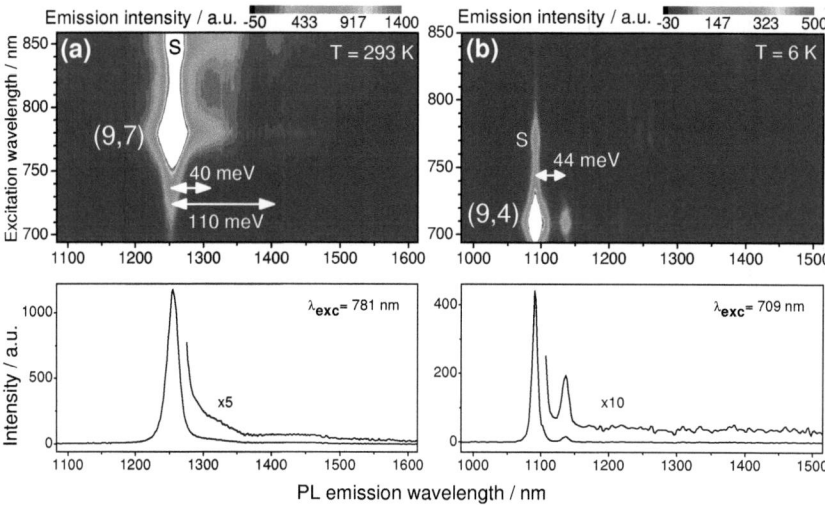

Figure 5.10: µ-PLE contour maps from suspended, single emitters at room temperature and at $T = 6\,\text{K}$. The bottom panels show PL emission spectra excited at the E_{22} absorption maxima of the nanotubes. PLE sidebands are labeled with 'S' and the assignment corresponds to that of Sec. 5.1. **(a)** $(9,7)$ SWNT at ambient temperature. Two weak emission satellites are visible. **(b)** $(9,4)$ SWNT at 6 K. Only one emission satellite is observed with a shift similar to that in **(a)**. From Ref. [11].

sufficiently high signal-to-noise ratio at room temperature. We also measured three contour maps at temperatures between 5-7 K. Due to the reduced linewidth compared to ambient temperature, we observed well-resolved but somewhat weaker DE_1-satellites at $\sim 40\,\text{meV}$ and very weak or no features at lower energies. Figure 5.10(b) shows a $(9,4)$ SWNT at 6 K and a corresponding PL spectrum at maximum absorption. Altogether, two groups of emission satellites were found for seven different (n_1, n_2) chiralities $[(8,7); (9,4); (9,7); (10,2); (11,3); (12,1); (12,4)]$, redshifted by $\Delta E_{DE_1} = 38 - 45$ and $\Delta E_{DE_2} = 100 - 130\,\text{meV}$ relative to OE_1, correspondingly. The intensities of the satellites were between 30 and 300 times lower than those of the main PLE peaks.

Apart from µ-PLE spectra of individual, air-suspended SWNTs, we also observed the above-mentioned satellites upon OE_1 excitation in water-surfactant dispersions. The

99

ensemble PLE contour map in Fig. 5.11(a) shows HiPco SWNTs dispersed in D_2O/SDS excited in the OE_1 region (~ 900-1500 nm). Typical features include the main OE_1 excitation/OE_1 emission PL signals, superimposed with Rayleigh scattering as well as exciton-phonon complexes of the OE_1 exciton with G- and D^*-mode phonons. All of these features are well-known in the literature [2, 52, 78, 83]. There is another group of PLE signals, indicated by a red dashed line and labeled with DE_2. A combined inspection of these peaks and the OE_2 region (not shown) suggests that these features cannot be assigned to any known PLE features. They correspond to OE_1 excitation and $(OE_1 - \Delta E)$ emission as indicated for $(9,5)$ nanotubes in Fig. 5.11(a). The same emission satellites can be even better seen in the PLE map of Fig. 5.11(b), where the $(8,7)$ SWNT was enriched by ultracentrifugation. This map also shows the OE_2 excitation/OE_1 emission range which indicates that $(8,7)$ is the most prominent nanotube in the ensemble. Consequently, the emission satellite corresponding to this tube at OE_1 excitation is most prominent as well and can be clearly assigned to $(8,7) - DE_2$ in Fig. 5.11(b). A corresponding feature is also expected for OE_2 excitation (as in Fig. 5.10) and a faint peak is indeed observed—albeit weaker than that excited at OE_1. The PLE contour map in Fig. 5.12 shows a $(9,7)$ enriched carbon nanotube dispersion where DE_2 satellites excited at both OE_1 and OE_2 are more distinctively observed. Therefore, the redshift of the emission satellites for a $(9,7)$ SWNT is about 125 meV. In contrast to air-suspended SWNTs, the intensity of the DE_2 emission satellites in ensembles was only ~ 10 times less than the main excitonic OE_2 excitation/OE_1 emission transition. The stronger satellites are likely due to stronger perturbations experienced by dispersed SWNTs.

Ensemble PLE contour maps of other 'standard' (i.e. not enriched) HiPco and PLV nanotubes dispersed in SDS or SDBS show similar emission satellites (acquired from OE_1 excitation due to aforementioned reasons) redshifted by $\Delta E_{DE_2} = 95 - 143$ meV. Figure 5.13(a) plots ΔE_{DE_2} as a function of nanotube diameter. A significant increase for tubes above $d = 0.85$ nm can be observed. It is possible that dispersed SWNTs also show DE_1 emission satellites. However, their detection was hindered due to strong Rayleigh scattering and broad ensemble PLE peaks (see Figs. 5.11 and 5.12). The two groups of emission satellites observed for air-suspended and water-dispersed SWNTs— DE_1 and DE_2—are tentatively attributed to dark excitonic states below the first bright OE_1 exciton. We rule out phonon-assisted transitions as a possible explanation because of the observed diameter dependence shown in Fig. 5.13(a). The schematic diagram of

5.3 Direct Observation of Deep Excitonic States in the PLE Spectra of SWNTs

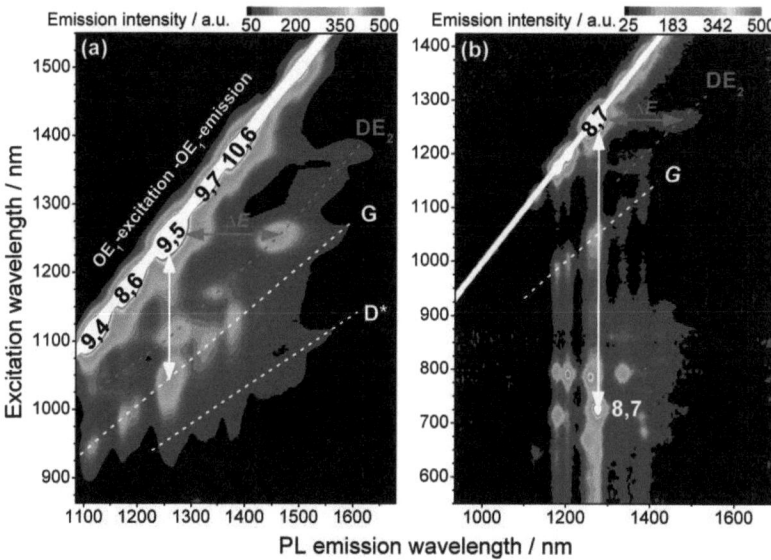

Figure 5.11: PLE contour maps from nanotube ensembles in dispersion. (a) HiPco nanotubes dispersed in $D_2O/1$ wt% SDS. The white, sharp stripe of high intensity is due to Rayleigh scattered excitation light. Broadening of this stripe is a result of different indicated nanotubes and their OE_1 PL emission which is almost identical to OE_1 excitation since the Stokes shift is small. The assignment of the broadenings is determined from the characteristic OE_2 excitation region between 550 and 900 nm (not shown). The white G and D^* lines denote $(OE_1 + G)$ and $(OE_1 + D^*)$ phonon-assisted excitations and OE_1 emissions. The red DE_2 line indicates emission peaks from low-energy dark excitonic states excited via OE_1. Arrows correlate phonon-assisted and dark excitonic peaks with a (9,5) SWNT. (b) PLV nanotubes dispersed in $D_2O/1$ wt% SDBS, enriched in (8,7) SWNTs by density gradient ultracentrifugation. The contour map combines both OE_2 (∼ 600-900 nm) and OE_1 (∼ 1100-1400 nm) excitation regions. Arrows link the zero-phonon peaks of (8,7) with DE_2 and dashed lines have the same meaning as in (a). Note that the DE_2 peak is most prominent and unobscured in this sample. From Ref. [11].

energy levels in semiconducting SWNTs in Fig. 5.13(b) summarizes our results. DE_1 and DE_2 are approximately 40 and between 100 and 140 meV below OE_1, respectively and can be populated via excitation of OE_1 or OE_2, followed by nonradiative relaxation.

5 Spectroscopic Characterization of Individual Carbon Nanotubes

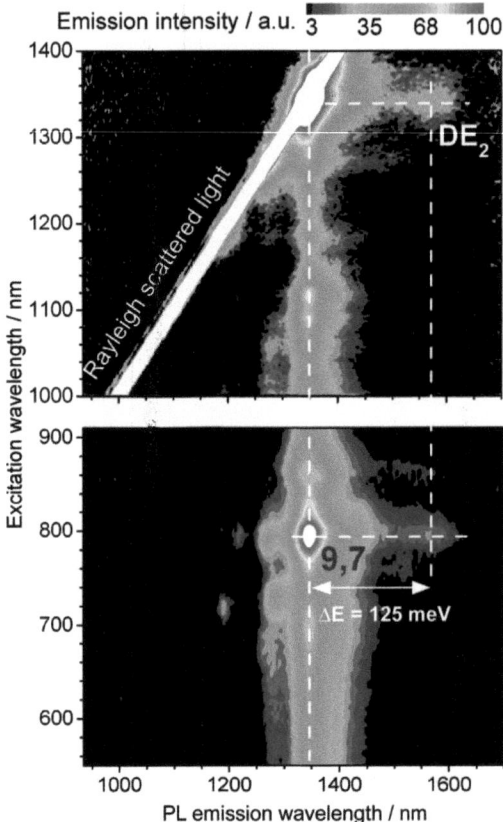

Figure 5.12: PLE contour map from PLV nanotubes dispersed in $D_2O/1$ wt% SDBS, enriched in (9,7) SWNTs by density gradient ultracentrifugation [33]. Both OE_2 and OE_1 excitation regions are shown. Dashed lines link OE_2 excitation/OE_1 emission, OE_1 excitation/OE_1 emission and corresponding DE_2 emission satellites of the dominant (9,7) nanotubes. From Ref. [11].

Surprisingly, the energy separation ΔE_{DE_2} for tubes of the same chirality does not change much for air-suspended or dispersed SWNTs. On the other hand, the binding energy of the OE_1 exciton is known to decrease with increasing dielectric screening [9, 56, 64, 66, 150, 188] and Jiang et al. have recently predicted an increase in ΔE for the lowest-energy dark exciton of a (10, 0) SWNT when taking dielectric screening into account [56]. Triplet states are not included in the scheme of Fig. 5.13(b), although they have been considered theoretically [56, 73, 75, 77]. However, we believe that they are not significant for the photophysics of SWNTs due to the weak spin-orbit coupling

5.3 Direct Observation of Deep Excitonic States in the PLE Spectra of SWNTs

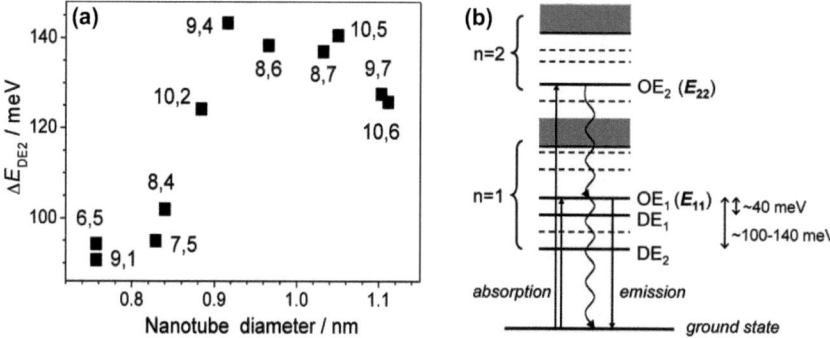

Figure 5.13: (a) Energy difference between OE_1 and low-energy dark excitonic states DE_2, ΔE_{DE_2}, as a function of nanotube diameter d. The shifts have been extracted from various PLE maps of HiPco and PLV nanotubes dispersed in D_2O/SDS or SDBS and excited in the OE_1 spectral region. Experimental error bars are ± 5 meV. (b) Schematic diagram of excitonic manifolds (horizontal lines) and electron-hole continuum bands (shaded areas) in semiconducting SWNTs. One-photon optically allowed states and transitions into/from OE_1 and OE_2 is indicated by vertical arrows. Radiationless decay is depicted by wiggled arrows. The presence of possible excitonic states other than DE_1 and DE_2 (lying ~ 40 meV and $\sim 100\text{-}140$ meV below OE_1) which were not detected in this study is indicated by dashed lines. From Ref. [11].

of carbon, at least if spin interconversion due to external perturbations (e. g. catalyst particles) can be neglected.

In our very recent studies on dispersions of SWNTs in toluene, highly enriched with a single chirality (so-called 'monodispersions'), the existence of two weakly allowed emitting states below OE_1 was confirmed and their effect on quantum yield and relaxation pathways was further elucidated [35]. These studies showed that the majority of OE_2 excitons relax first to OE_1. It appears, however, that DE_1 and DE_2 can be directly populated from OE_2 via a minor relaxation channel ($\lesssim 20\%$). In the future, it will be of interest to probe modifications of such deep dark states by controlled mechanical deformation or the application of external electric or magnetic fields.

5.4 Spectroscopy and AFM Manipulation of Individual, Ultralong Carbon Nanotubes

5.4.1 Background

SWNTs in typical dispersions exhibit average lengths of a few hundred nanometers as determined by AFM imaging of deposited (e.g. spin-coated) nanotubes [145]. Even if the starting material consists of very long tubes, e.g. VANTAs, tip-ultrasonication efficiently cuts them into short pieces. On the other hand, (far-field) PL and Raman microscopes have an optimum resolution of 300 nm, so SWNTs deposited on a surface from dispersion can only be probed 2-3 times going from one end to the other. Therefore, we tried to grow individual, long SWNTs by CVD to observe their PL properties along the z-axis and to check structural integrity (changes of chirality) along a nanotube. However, as opposed to HiPco and PLV SWNTs, nanotubes grown by thermal CVD with a supported catalyst tend to have diameters larger than 1.4 nm which corresponds to SWNTs emitting PL above 1600 nm ($\cong$ the upper limit of our InGaAs detector). Additionally, Lefebvre et al. and Jeong et al. claimed that the PL of individual, as-grown SWNTs is completely quenched when being in direct contact with a surface [189, 190] and that only air-suspended sections of SWNTs show PL (or surfactant-coated when deposited from dispersion). In Sec. 5.2, we have already reported on the possibility to observe PL from SWNTs in contact with a surface, albeit with very low signal intensities of only ~ 5 % compared to air-suspended tubes. In this work and in Ref. [12, 191], we discuss PLE and Raman spectra along ultralong SWNTs and perform AFM and SEM imaging as well as AFM manipulation on the same nanotube. The following samples of individual, surface-bound SWNTs were used for this study:

- Horizontally aligned carbon nanotube arrays from CO-CVD on sapphire (Fig. 4.9)

- Long, individual SWNTs from ethanol CVD on Si/SiO_2 (Fig. 4.8)

- SWNTs synthesized by methane CVD from Li et al. on prepatterned Si/SiO_2 (Fig. 4.11) [136].

The samples received from Li et al. contained etched trenches and markers which allowed for measurements of air-suspended SWNT sections and also facilitated tube retrieval via SEM, AFM and PL microscope. Hence, most of this section focuses on these samples.

5.4.2 Nanotubes from CO-CVD

The ultralong nanotubes shown in Fig. 4.9(a) and (b) seem to be ideal samples for PLE mapping and PL imaging along the tube z-axis. However, none of these tubes protruding over several hundred micrometers displayed measurable PL. Only close to the edge between catalyst pad and substrate, PL signals from multiple SWNTs were observed. However, PL imaging (with an excitation power of 3-4 mW, an excitation wavelength of 830 nm, a piezo step size of 500 nm and the Olympus 100x/0.95 objective) showed that most such nanotubes exhibited rather localized, 'point-like' PL as depicted in Fig. 5.14(a) as opposed to elongated, 'tube-like' images as shown in Fig. 5.14(b). In fact, Fig. 5.14(b) was the only observable PL image of this sample which indicated a straight and long SWNT resting completely on the sapphire surface. However, finding this spot again was not possible due to a lack of markers. The insulating substrate also prevented high-magnification SEM images.

We therefore focused on the investigation of the very long nanotubes visible in the SEM images to find out why they were not showing any measurable PL. For that purpose, we used Raman spectroscopy, Raman imaging (with 633 nm excitation, 2-3 mW excitation power and a 100x/0.9 objective) and AFM [Veeco Instruments, 'Multi Mode' with NSC15 silicon cantilevers (MikroMasch) in intermittent contact mode]. Figure 5.15 shows a few exemplary measurements of 4 long nanotubes. Figure 5.15(a) is a SEM image which facilitated orientation on the sample. Horizontal, black bars mark regions of the tubes, where AFM images and height profiles were recorded. Rectangles denote areas where Raman imaging was performed. Two of them are depicted in Fig. 5.15(b) and (c) (10x20 µm and 50x50 µm, respectively). They correspond to the red and green rectangle, respectively and were measured with vertically polarized excitation light. Thus, horizontal sections of the tubes in both images show less intensity than vertical ones. The step size in Fig. 5.15(b) was 0.2 µm, and for image build-up, the area around the G-mode was integrated. No RBMs were detected[3] and average AFM heights were 1.3 ± 0.3 nm. The image in Fig. 5.15(c) was recorded with a step size of 0.4 µm and integration was performed over the D^*-mode, thereby visualizing all three strands of tube (2) and (3). However, tube (2) also exhibited 3 RBMs corresponding to diameters of 1.7, 1.9 and 2.3 ± 0.1 nm as shown in Fig. 5.15(d). The error is mainly due

---

[3]The resonance window for SWNTs in RRS is approximately $E_{ii} \pm \omega_{mode}$. If 633 nm is not within $E_{ii} \pm \omega_{RBM}$, it can still be within $E_{ii} \pm \omega_G$ as $\omega_{RBM} < \omega_G$.

5 Spectroscopic Characterization of Individual Carbon Nanotubes

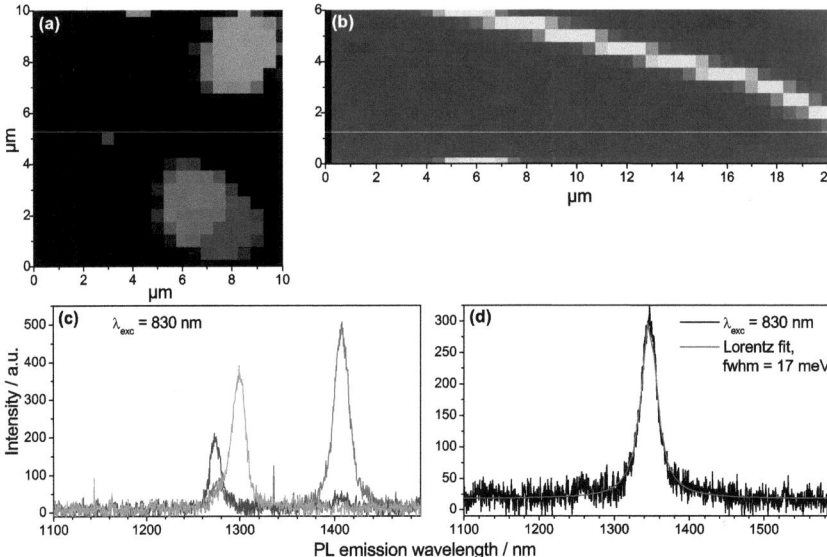

Figure 5.14: PL images of individual, surface-bound SWNTs grown by CO-CVD, close to the catalyst pad. Excitation wavelength and piezo step size (≙ pixel size) is 830 nm and 500 nm in (a) and in (b), respectively. Images were acquired by integrating over specific areas in the spectrum (see Fig. 3.3 on page 50). The catalyst pad is close to the right of each image. Laser excitation light is horizontally polarized. (a) Typical 'point-like' image of three different nanotube species, represented by blue, green and red. Overlapping colors (e.g. the red and blue areas) represent spatially close (but spectrally separated) SWNTs. Individual spectra from the center of each nanotube 'spot' are shown in (c). (b) Individual, elongated SWNT. (d) PL emission spectrum from the image depicted in (b). A Lorentz fit (in red) exhibits a typical fwhm of 17 meV.

to an assumed ∼5% uncertainty of the proportionality constant $A = 248\,\text{cm}^{-1}\text{nm}$ [88]. The largest spectroscopically determined diameter corresponds well to average AFM heights of 2.5 ± 0.2 nm. We believe that tube (2) is an example of a small bundle of SWNTs. We rule out the existence of a triple-walled carbon nanotube (TWNT), since the differences in diameters do not correspond to 0.72 ± 0.02 nm as measured by X-ray diffraction of double-walled carbon nanotubes (DWNTs) [192]. However, a bundle of

5.4 Spectroscopy and AFM Manipulation of Individual, Ultralong Carbon Nanotubes

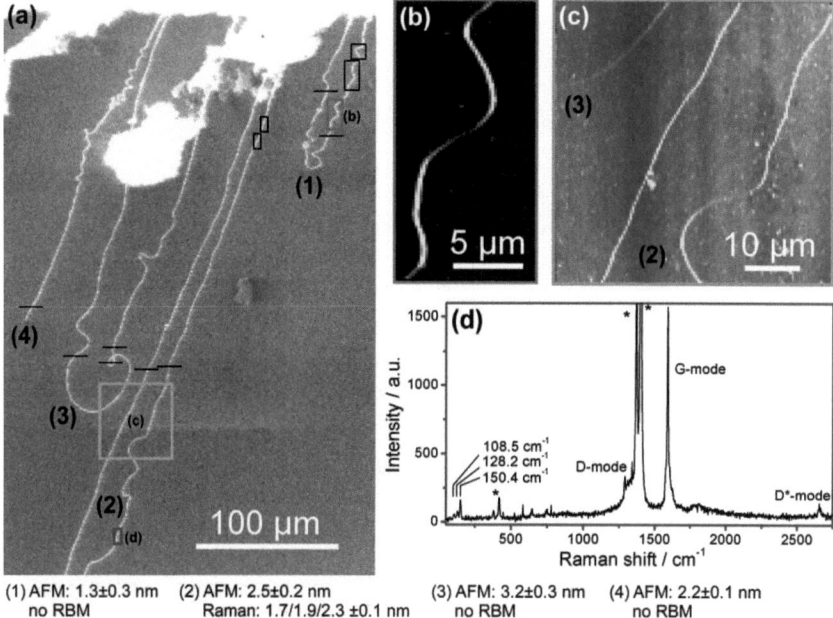

(1) AFM: 1.3±0.3 nm   (2) AFM: 2.5±0.2 nm   (3) AFM: 3.2±0.3 nm   (4) AFM: 2.2±0.1 nm
no RBM   Raman: 1.7/1.9/2.3 ±0.1 nm   no RBM   no RBM

Figure 5.15: SEM and Raman images/spectra of very long, surface-bound nanotubes grown by CO-CVD, far away from the catalyst pad. (a) SEM image serving as a 'map' for Raman spectroscopy and AFM imaging. Horizontal, black bars indicate AFM height measurements. Rectangles mark positions of Raman images. Two such images and one Raman spectrum are depicted in (b), (c) and (d), respectively. (b) Raman image of the red area of SWNT (1) (see text for details). (c) Raman image of the green area, showing SWNT (2) and (3) (see text for details). (d) Single Raman spectrum of the nanotube within the blue rectangle shown in (a). Three RBM peaks are visible. Background Raman signals from the sapphire substrate are denoted with an asterisk. Diameters determined by AFM or Raman for all 4 visible nanotubes are listed below.

one SWNT (1.9 nm) and one DWNT (1.7 nm as inner and 2.3 nm as outer tube) might be possible. Whatever scenario is true, it shows that small bundles of nanotubes might be confused with individuals when considering AFM height measurements alone. It also proves that a simultaneous growth of several nanotubes in bundled form is possible. Tube (3) was only weakly in resonance with 633 nm excitation light (only a weak D^*-mode was observed) and no Raman signal was recorded for tube (4).

In total, we performed µ-Raman spectroscopy on 20 nanotubes. For 11 tubes we were able to determine RBM frequencies and thus diameters which ranged between 0.5 and 2.5 nm, with an average of 1.8 ± 0.5 nm. 8 tubes exhibited several (either two or three) RBM signals. Only for 2-3 tubes did the spacing of diameters agree with that of a DWNT. About 70 individual AFM height measurements revealed diameters between 0.6 and 3.7 nm, with an average of 2.1 ± 0.8 nm. There was no nanotube that showed both only one RBM signal and a RBM frequency below 1.4 nm (our largest detectable diameter by PL). We therefore believe that the diameter of most nanotubes was too large to be detected via our PL set-up, and that the small amount of nanotubes with $d \leq 1.4$ nm (about 5 RBM frequencies and 2 AFM measurements indicating such a diameter) was either bundled, an inner tube of a DWNT, metallic or showing PL too weak to be detected.

5.4.3 Nanotubes from Ethanol CVD

The bulk Raman spectrum in the inset of Fig. 4.8(d) indicates SWNTs with diameters small enough to be detected via PL. Loop structures like the one shown in the inset of Fig. 4.8(e) show interesting electroluminescent effects in the field of carbon nanotube optoelectronics [193]. However, we could not find any luminescent SWNTs exhibiting loop structures. Figure 5.16 shows SEM, AFM and Raman images of the same loop in a SWNT of 1.6 nm diameter as determined via AFM. The tube was made by ethanol CVD, rests on SiO_2 and is part of a SWNT with a length of more than 100 µm. By correlating scanning electron with optical micrographs, the loop could be retrieved in the different instruments. The G-mode Raman images in Fig. 5.16(c) and (d) are both identical (633 nm excitation, excitation power of ~ 2 mW, 10×10 µm with a step size of 0.1 µm, 100x/0.9 objective) except for the polarization. In Fig. 5.16(a)[(b)] the excitation light is horizontally [vertically] polarized, amplifying only those sections of the SWNT which are aligned parallel to the polarization. The intensity I is thereby related to the angle α between tube z-axis and polarization vector $\vec{P}$ via [194]

$$I \propto \cos^2 \alpha. \tag{5.2}$$

5.4 Spectroscopy and AFM Manipulation of Individual, Ultralong Carbon Nanotubes

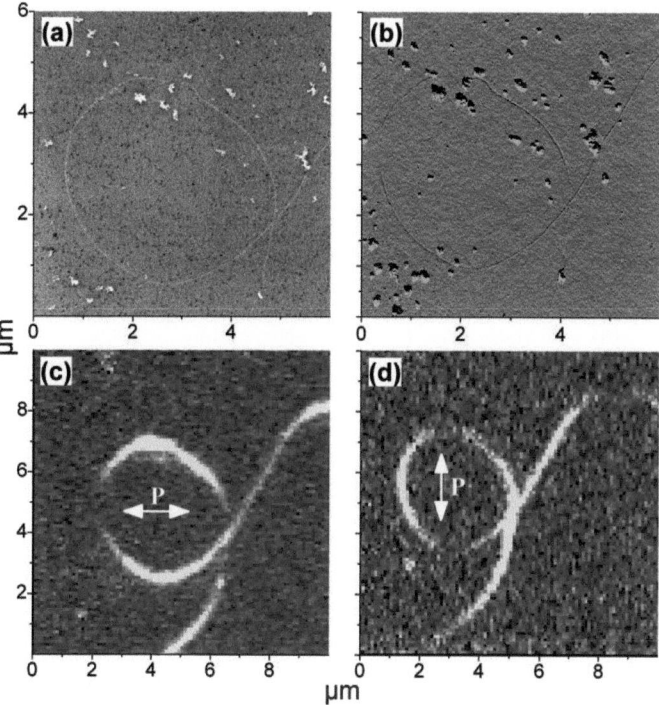

Figure 5.16: (a) SEM, (b) AFM error signal, (c) and (d) Raman images of the same loop in a SWNT grown by ethanol CVD. $\vec{P}$ denotes the polarization vector of the excitation light (see text for details).

The polarization of excitation light in our Raman microscope cannot be changed due to a highly polarizing holographic beamsplitter cube [(6) in Fig. 3.4 on page 52]. Therefore, the sample was rotated by 90° to acquire the Raman image of Fig. 5.16(d).

Inside an indentation of the catalyst pad, we were also able to find an individual, 50-µm-long SWNT on this sample showing PL [191]. An SEM image of this tube (combined of several individual ones, with 10 kV acceleration voltage) is depicted in Fig. 5.17(a). The contrast was strongly enhanced for an optimized view of the tube, thereby overexposing the catalyst edges. In the optical bright field micrograph of

5 Spectroscopic Characterization of Individual Carbon Nanotubes

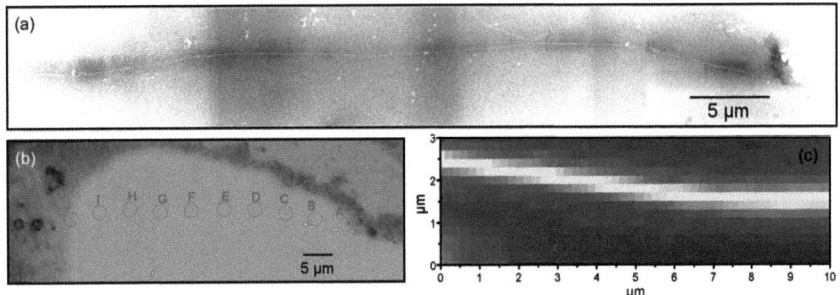

Figure 5.17: (a) SEM image assembled from 5 overlapping SEM pictures showing a 50-μm-long section of a SWNT. (b) Optical bright field image of the same area as in (a). Circles (A)-(K) follow the SWNT and denote positions of measured PLE contour maps (see Fig. 5.18). (c) PL image of a 10-μm-long section of the same nanotube gathered by exciting the SWNT at 850 nm and integrating the PL emission signal around 1480 nm. From Ref. [191].

Fig. 5.17(b), the catalyst 'fjord' which is crossed by the SWNT can be seen much better. Circles mark the positions of measured PLE contour maps, each shown in Fig. 5.18(A)-(K). A PL image of a 10-μm-long section of the same SWNT is illustrated in Fig. 5.17(c). The PL spectra for the image were acquired at an excitation wavelength of 850 nm, an excitation power of ∼5 mW, a step size of 0.2 μm and with the Leica oil immersion objective (100x/1.4) described in Tab. 3.1. The PL image was then assembled by integrating each spectrum around 1480 nm. Due to the small diameter of the SWNT, it can be essentially regarded as an infinitesimally thin, luminescent rod. The fwhm of a vertical cross section of the nanotube PL image is therefore a measure for the resolution of the microscope. On average, we thus determined a resolution of about 400 nm which is close to the theoretically determined resolution of the objective of about 300 nm. Additionally, no significant PL intensity changes along the SWNT can be seen (variations are within ±5%), i.e. light is emitted uniformly on a length scale larger than 400 nm. The polarization vector of the incident beam is approximately parallel to the tube axis in Fig. 5.17 and for all PLE maps in Fig. 5.18.

The PLE contour maps of Fig. 5.18(A)-(K) were recorded at intervals of 5 μm. The excitation wavelength was tuned in steps of 2 nm. The spectra were cut on the low wavelength side (below 1300 nm) to block the large background luminescent signal of silicon.

5.4 Spectroscopy and AFM Manipulation of Individual, Ultralong Carbon Nanotubes

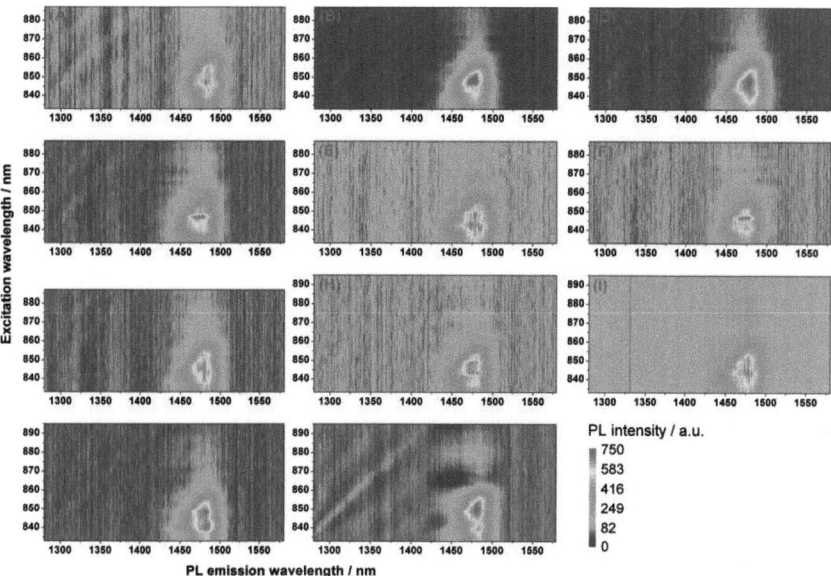

Figure 5.18: PLE contour maps recorded at eleven 5-µm-spaced positions (A)-(K) shown in Fig. 5.17(b). From Ref. [191].

A close inspection of the signals (via fitting of Lorentzian functions in excitation and emission) shows that the PL maximum (fwhm) varies between 842-847 nm (18-20 meV) in excitation and between 1472-1481 nm (40-48 meV) in emission. Average values are 844 ± 1 nm (fwhm: 18.5 ± 0.1 meV) and 1476 ± 3 nm (fwhm: 44.8 ± 0.3 meV), respectively. A PLE contour map with a larger range of excitation wavelengths is depicted in Fig. 5.19(a). We did not observe any other signal within our detection range. We attribute the small variations in emission/excitation energies to changes in the local surroundings of the tube and to its interaction with the substrate. The oblique stripes in the PLE contour maps of Fig. 5.18 and 5.19(a) are probably due to higher order Raman signals, either from the tube itself or from impurities (e.g. soot). One of these Raman signals seems to overlap with the main electronic transition and to produce an asymmetric peak shape. Despite this effect, Fig. 5.18 and 5.19(a) clearly show that the SWNT retains its chirality at all probed spots along 50 µm. The uniform structure

5 Spectroscopic Characterization of Individual Carbon Nanotubes

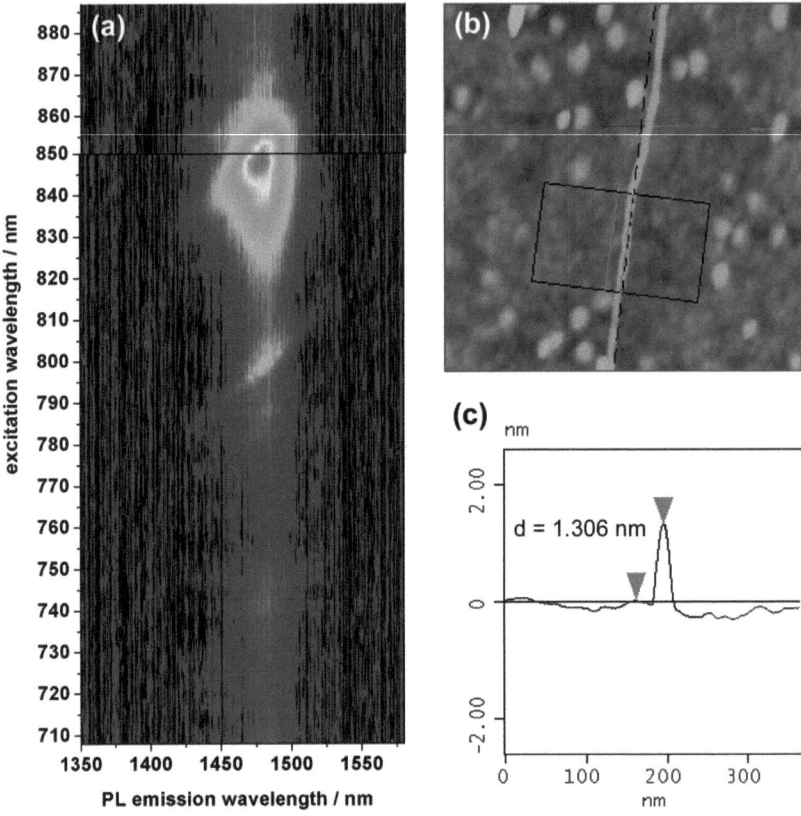

Figure 5.19: (a) PLE contour map at position (C) in Fig. 5.17 over a larger range of excitation wavelengths (708-887 nm), recorded with Ti:sapphire laser I and II. (b) AFM image of the nanotube. The rectangle marks a region which is used to calculate an average profile shown in (c). The vertical distance of the red markers in (b) and (c) corresponds to the diameter of the SWNT. The value of 1.306 nm agrees well with that of a (10, 9) SWNT (1.29 nm).

of the nanotube likely indicates a uniform growth process. Comparison to the PL of nanotubes in D_2O/surfactant [100] and to PL of air-suspended SWNTs (Sec. 5.1, [9]) suggests that this tube's chiral indices are (10, 9) (see also Fig. 5.3(b) on page 78 where

5.4 Spectroscopy and AFM Manipulation of Individual, Ultralong Carbon Nanotubes

λ_{11} and λ_{22} of this tube were included). This assignment is further confirmed by AFM height measurements of this SWNT. Figure 5.19(b) shows an AFM image and the corresponding section used for the calculation of an average profile [Fig. 5.19(c)]. The height is determined to 1.306 nm which is in very good agreement with the theoretical value of 1.29 nm for a (10, 9) tube.

5.4.4 Ultralong and Aligned Nanotubes from Li *et al.*

The CVD synthesis of the nanotubes that were obtained from Li *et al.* in Beijing (collaborative project) has already been discussed in Sec. 4.2. The exemplary SEM image in Fig. 4.11 shows the markers, the trenches and the alignment of the SWNTs growing over them. By comparing PLE spectra of suspended and unsuspended parts of the same SWNT, we were able to quantify for the first time changes in (i) quantum yield, (ii) E_{11}/E_{22} energies and (iii) line widths of E_{11} emission and E_{22} excitation. Afterwards, homogeneous parts of SWNTs on the surface were selected and manipulated by atomic force microscopy (AFM), thereby introducing torsional and uniaxial strain as well as defects, similar to Duan *et al.* [195]. In contrast to Duan *et al.* who studied the effect of AFM manipulation via RRS, we use PLE spectroscopy which is believed to be much more sensitive to structural changes of the tube.

The horizontally aligned SWNT arrays were grown on a photolithographically prepatterned silicon substrate with a 500 nm thick oxide layer [136]. The layout is sketched in Fig. 5.20 and consists of a parallel array of 3.5 µm wide, 500 nm deep and several millimeters long trenches, oriented perpendicular to the growth direction. The trenches have a spacing of 300 µm and are marked with etched numbers and letters (height x width $\approx$ 40 x 20 µm, line thickness of 5 µm and the same depth as the trenches) to facilitate tube retrieval. These markers partially served to suspend SWNTs, as well.

The detection window for PL emission was set between $\sim$ 1200 nm and 1600 nm, taking into account the background luminescence of the Si substrate and the upper limit of our InGaAs detector, respectively. This translates into a diameter range between $\sim$ 1.1 and 1.4 nm which is at the lower end of the SWNT diameter distribution of this sample (between 0.9 and 3.3 nm with a mean value of 2 nm, as measured by AFM [136]). It could explain why only $\sim$1 out of 10 SWNTs bridging the trenches and being visible in SEM could be detected by PL. Assuming a 2/3 fraction of semiconducting SWNTs and no 'missed' tubes, only 15% of all semiconducting SWNTs were detected. This fits well

5 Spectroscopic Characterization of Individual Carbon Nanotubes

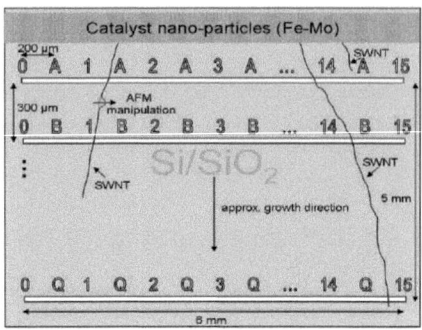

Figure 5.20: Sample layout of ultralong, horizontally aligned SWNT arrays from Li *et al.* Depth of trenches and markers is about 500 nm. The red section of the left SWNT shows how we manipulated it by AFM.

to a Gaussian distribution of diameters centered at 2 nm with a fwhm of 1.2 nm.

PLE spectra were typically acquired with excitation powers of 1-2 mW and fitted with a 2D Lorentz function, yielding the position of the maximum, fwhm$_{E11}$ in emission, fwhm$_{E22}$ in excitation and the 'volume' which was corrected (divided) by the excitation power and the acquisition time and taken as a measure for the intensity. Typical acquisition times ranged between 10 and 15 s for a PL spectrum of a surface-bound SWNT and between 2 and 5 s for air-suspended ones. All PLE maps and PL images were recorded with the 100x/0.95 Olympus objective.

Results of unmanipulated SWNTs Figure 5.21 shows E_{ii}, fwhm$_{Eii}$ ($i = 1, 2$) and intensities acquired for 3 different individual SWNTs as a function of position. Based on results discussed in Ref. [9] measured E_{11}-E_{22} values fit best to chiral indices (15, 2) or (16, 0) for SWNT 1, (13, 2) or (14, 0) for SWNT 2 and (12, 4) for SWNT 3 [the spectral positions of both air-suspended and surface-bound sections were also included in Fig. 5.3(b)]. Suspended sections are marked with a gray shading. For all other positions the SWNT rests directly on SiO$_2$. Assignment, shifts of E_{ii}, average values of fwhm$_{Eii}$ ($i = 1, 2$) and intensity drops for 4 different SWNTs are summarized in Tab.5.2. Values for ΔE_{ii} are positive when going from surface-bound to air-suspended sections of a nanotube. The E_{22} maximum of SWNT 4 is beyond our excitation range (> 988 nm) so only ΔE_{11} and fwhm$_{E11}$ values are given in Tab. 5.2. A clear assignment of chirality was not possible.

For SWNT 1, Pos. 0 corresponds to the edge of the trench [see also Fig. 5.22(a)]. This explains why the values of E_{11}, E_{22} and the intensity are in between those of suspended and unsuspended tubes. Pos. 250-265 corresponds to parts of SWNT 1, freely suspended

5.4 Spectroscopy and AFM Manipulation of Individual, Ultralong Carbon Nanotubes

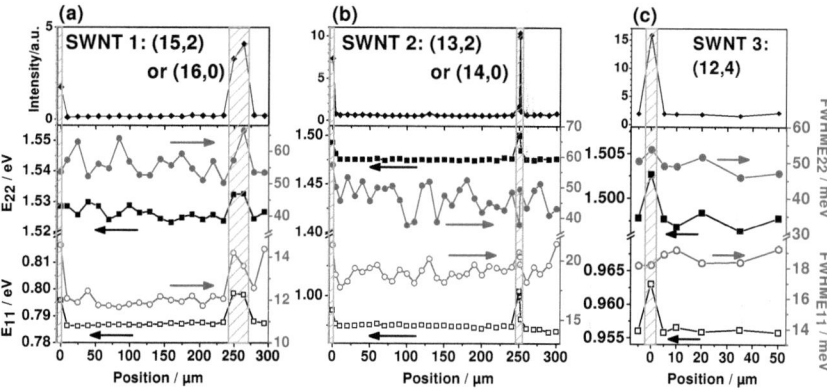

Figure 5.21: E_{ii}, fwhm$_{Eii}$ ($i = 1, 2$) and intensities acquired for 3 different individual SWNTs as a function of position. Grey shadings mark trenches or etched markers where tubes are suspended. Air-suspended sections show both an increase in E_{ii} and emission intensity. Changes in the fwhm are inconclusive (see text for details).

Table 5.2: Summary of measured values for 4 different SWNTs. The E_{22} maximum of SWNT 4 is beyond our range (> 988 nm). A clear assignment is therefore not possible.

SWNT No.	(n_1, n_2)	$(n-m)$ mod 3	ΔE_{11} (meV)	ΔE_{22} (meV)	$\langle fwhm_{E11} \rangle$ (meV)	$\langle fwhm_{E22} \rangle$ (meV)	I_s/I_f %
1	(15,2)/(16,0)	+1	11	8	14.1 ± 0.5	56 ± 5	5-8
2	(13,2)/(14,0)	-1	27	24	19.2 ± 1	47 ± 5	6
3	(12,4)	-1	7	5	18.6 ± 0.5	49 ± 3	12
4	(16,2)?	-1?	23	-	16.5 ± 2	-	7

over etched markers. The maximum blue-shift due to suspension is 11 and 8 meV for E_{11} and E_{22}, respectively. This corresponds to 1.4 and 0.4 %, referred to values of unsuspended tubes. fwhm$_{E11}$ seems to increase by ~2-3 meV at the suspended parts of SWNT 1. However, we attribute this to an artifact, because the emission of the unsuspended tube is in a spectral range (close to 0.775 eV) where the quantum efficiency of our detector drops dramatically. The peak is therefore cut-off at its low energy side and the fitting produces a smaller line width. This is not the case for a suspended

5 Spectroscopic Characterization of Individual Carbon Nanotubes

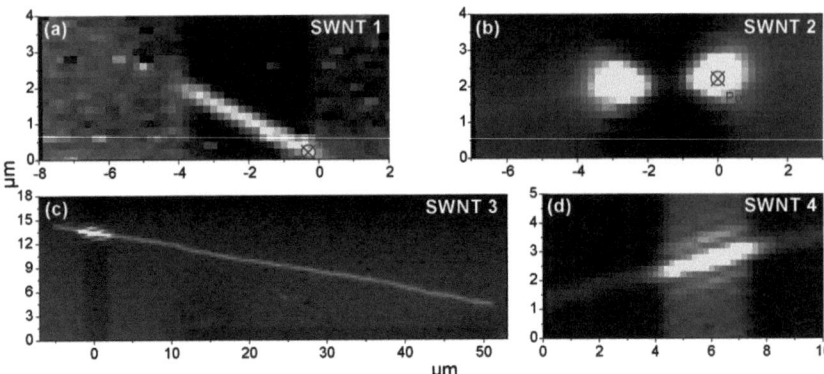

Figure 5.22: PL images of SWNT 1-4. Vertical, dark (except for image of SWNT 4) stripes correspond to the trench. **(a)** SWNT 1 continues on the surface but acquisition time was too short. Pos. 0 marks the beginning of the diagram in Fig. 5.21(a). **(b)** SWNT 2 touches the bottom in the middle of the trench and is therefore too dim to be seen, the same holds for the surface. Again, the blue cross marks the beginning of the diagram in Fig. 5.21(b). **(c)** PL image of the 55-µm-long section of SWNT 3. The x-axis corresponds to the one in Fig. 5.21(c). The tube continues in both directions beyond the image without showing PL. **(d)** PL image of SWNT 4. An increased background offset in the trench reverses the color as compared to the other images.

tube with blue shifted emission. However, this should not affect fwhm$_{E22}$. The average value of fwhm$_{E22}$ is 56±5 meV and 14.1±0.5 meV for fwhm$_{E11}$. Line widths do not show a significant change due to suspension within the mentioned error bars, apart from aforementioned deficiencies. However, if a similar percentual change were to pertain to the fwhm as observed for E_{11} and E_{22}, error bars would presently still be too large to make a final statement.

For SWNT 2, the tube touches the bottom of the trench close to Pos. 0 as shown in the PL image of Fig. 5.22(a), so only ∼1 µm is suspended. As with SWNT 1, E_{11}, E_{22} and intensity are not as high as around Pos. 250 with a suspended section of ∼5 µm.

SWNT 3 is a tube which probably changes chirality along its length. Therefore, only a 55-µm-long section could be measured. Before and after this section, the SWNT continues as seen with SEM. High excitation powers and long acquisition times did

5.4 Spectroscopy and AFM Manipulation of Individual, Ultralong Carbon Nanotubes

not yield any PL signal there. We therefore consider a very effective quenching in this region unlikely. The complete section is shown in Fig. 5.22(c). Pos. 0 corresponds to the suspended part of the tube in the middle of the trench.

All SWNTs measured show a blueshift between 7-27 meV and 5-24 meV for E_{11} and E_{22} when going from surface-bound to air-suspended sections, respectively. Referenced to water-surfactant dispersions, the blueshifts of surface-bound SWNTs are between 19-32 meV and 15-47 meV for E_{11} and E_{22}, respectively. ΔE_{11} energies agree well with those given in Sec. 5.2 for SWNTs on sapphire surfaces (15-30 meV and 10-25 meV for E_{11} and E_{22}, respectively) whereas ΔE_{22} is up to a factor of two larger [this is due to SWNT 3, see Fig. 5.3(b) where the results of the tubes measured in this section have also been included]. Experimental E_{ii} energies of suspended SWNTs are blueshifted compared to SWNTs surrounded by a medium because the former experience a lower dielectric screening (see Sec. 5.1). The E_{ii} energies of the surface-bound SWNTs (medium screening) are between those of the tubes in dispersion (highest screening) and those of air-suspended SWNTs [no screening, see Fig. 5.3(b) and Sec. 5.2]. SiO_2 has $\epsilon_r \approx 3.3$ at frequencies corresponding to a PL lifetime of $\sim$20-200 ps [196]. To a first approximation, ϵ_r^{eff} of a SiO_2 surface in vacuum becomes $\epsilon_r^{eff} = \frac{1}{2}(1 + 3.3) = 2.15$. This is an upper limit as the exciton exists in a nanotube which is located somewhat above the surface. This value is between that of surfactant-coated SWNTs in aqueous dispersions ($\epsilon_r = 2 - 3$, see Sec. 5.1) and air-suspended SWNTs ($\epsilon_r \approx 1$) which agrees with the observed energy shifts.

We see no effect of ϵ_r on fwhm$_{Eii}$ ($i = 1, 2$) within the error bars given in Tab. 5.2. Therefore, we only list average values. The scatter of the line width data is mostly due to additional Raman overtones close to or overlapping with the fundamental electronic transition (zero-phonon line), although we tried to compensate for it by subtraction. However, the line width is much more sensitive to such corrections than e. g. the center position. Inoue et al. have found that there is a diameter dependence of fwhm$_{E11}$ for air-suspended SWNTs [197]. They see a decrease of line width for increasing tube diameter. Their absolute values are by a factor of almost 2 smaller than ours, but we believe this is because they show only the lowest measured fwhm for each chirality. Nevertheless, we believe to see the same trend as Inoue et al. SWNT 2, 3 and 1 have diameters of 1.10-1.11, 1.13 and 1.25-1.26 nm and fwhm of 19.2, 18.6 and 14.1 meV. Only SWNT 4, having the largest diameter (1.4 nm), does not follow this trend. This could be due to an incorrect assignment or to the large scatter of the linewidth data. Hertel *et al.* have found a similar relationship between fwhm$_{E22}$ and E_{22} transition energy for SWNTs in SDS suspensions [198]. Their data suggests an increase of the line width with increasing E_{22} energy (decreasing λ_{22}). This is consistent with our data, including the (10, 9) tube from Fig. 5.19. Average values of λ_{22} (fwhm$_{E22}$) are 844 (45 meV), 840 (47 meV), 828 (49 meV) and 811 nm (56 meV) for (10, 9), SWNT 2, 3 and 1, respectively.

The intensity of the PL drops to 5-12 % for surface-bound SWNTs, referenced to suspended SWNTs. This explains why in former experiments of other groups the PL of SWNTs on surfaces was overseen. The number is also in good agreement with the ratio given on page 96. The involved quenching mechanisms which could explain why the quantum yield of surface-bound SWNTs decreases by a factor of 10-20 are still unknown. Clearly, all atoms of a carbon nanotube are surface atoms. It seems plausible to assume that the amount of channels for radiationless decay of excitons formed in SWNTs increases if a surface is brought into its vicinity.

Results for AFM manipulated SWNTs For AFM manipulation, we selected SWNT 1 and 2, because they belong to different $(n - m)$ mod 3-families and were well characterized over a length of 300 µm. Most importantly, they did not change their chiral indices within that distance. SWNTs in this sample typically grew over several mm and we confirmed that even after 600 µm, the chiral indices of SWNT 1 and 2 were still the same. We proceeded by choosing the middle position of the tubes between 2 trenches and performed AFM manipulation in contact mode via line-scanning along a path roughly perpendicular to the tube axis (at increased down force). Fig. 5.23 shows that SWNT 1 was thereby dragged away from its original position by almost 1 µm, whereas SWNT 2 was only moved by 200 nm. One can also see that both tubes were severely ruptured or squashed at the point of manipulation.

Figure 5.24 shows E_{ii}, intensity and fwhm$_{Eii}$ ($i = 1, 2$) as a function of position after manipulation for SWNT 1 [(15, 2) or (16, 0), $(n - m)$ mod 3 = +1]. Pos. 0 corresponds to the point of manipulation, values before manipulation are included, as well (in gray). The observation of additional PL shoulders indicated by two red dots in Fig. 5.24(a) is discussed below. Prior to manipulation of SWNT 1, a 13- µm-sized dirt particle was unintentionally dragged across the tube by the AFM cantilever, thereby introducing additional uniaxial strain. The dirt particle is between Pos. -32 and -45.

In Fig. 5.25, the same variables are plotted vs. position for SWNT 2 [(13, 2) or (14, 0),

5.4 Spectroscopy and AFM Manipulation of Individual, Ultralong Carbon Nanotubes

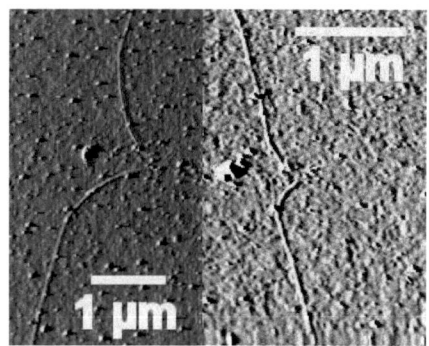

Figure 5.23: AFM error signal at the point of manipulation of SWNT 1 (left) and SWNT 2 (right). The manipulation was performed in contact mode via a line scan across the tube at increased down force.

$(n - m) \bmod 3 = -1]$ after (in red) and before (in gray) manipulation. Duan and coworkers have shown with Resonant Raman Spectroscopy [195, 199] that careful AFM manipulation introduces (i) uniaxial strain and (ii) also torsional strain as the tube rolls under the AFM tip due to friction between substrate and SWNT. We will first discuss torsion. Twisting a chiral nanotube with both ends fixed far away via rotation of its center is comparable to doing the same with a macroscopic steel spring (or a telephone cord): On one side the coil widens, on the other it narrows. This corresponds to torsional strain in different directions (with different signs) on the two sides. Theory therefore predicts a bandgap widening on one side, and a bandgap narrowing on the other [200, 201], of course superimposed by uniaxial strain if the rotation is performed by rolling on the surface. The effect of uniaxial and torsional strain is very similar regarding only the sign of ΔE_{ii}: (a) A change of sign in strain leads to a change of sign in ΔE_{ii} (b) the sign of ΔE_{ii} alternates with i (c) the sign of ΔE_{ii} changes depending on which $(n-m) \bmod 3$-family the nanotube belongs to. If we assume a certain handedness of the SWNT, we would expect an increase (decrease) of E_{11} (E_{22}) left from the point of manipulation and a decrease (increase) on the right. In addition, it should be exactly the opposite for a SWNT belonging to a different family, provided that handedness is equal.

Both SWNT 1 and SWNT 2 show a strong increase of E_{ii} ($i = 1, 2$) at Pos. 0 which cannot be attributed to torsional strain due to the aforementioned reasons (b) and (c). However, on the right-hand side of SWNT 1, where the superposition of uniaxial strain appears to be negligible, close inspection shows that for Pos. 0.5-3 ΔE_{11} is negative and

5 Spectroscopic Characterization of Individual Carbon Nanotubes

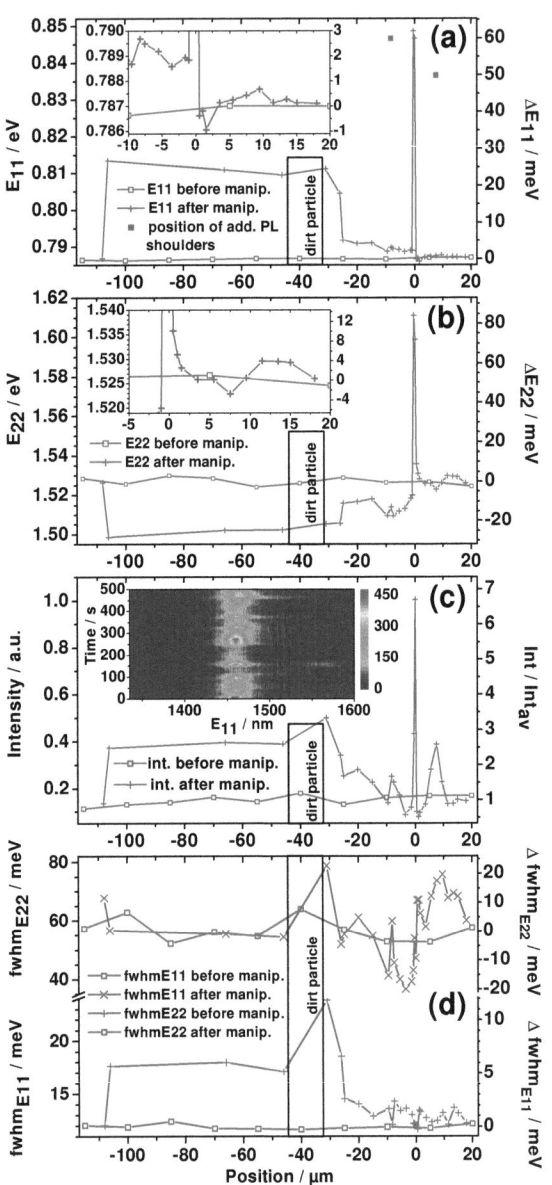

Figure 5.24: (a) E_{11}, (b) E_{22}, (c) intensity and (d) fwhm$_{Eii}$ ($i = 1, 2$) as a function of position before (gray) and after (red) manipulation of SWNT 1. Pos. 0 corresponds to the point of manipulation. The two red dots in (a) denote small PL emission shoulders about 8 μm away from the point of manipulation, with a spectral shift similar to that at Pos. 0. Insets in (a) and (b) show enlargements around Pos. 0. Inset in (c) is a time trace measured at Pos. -0.5, indicating blinking.

5.4 Spectroscopy and AFM Manipulation of Individual, Ultralong Carbon Nanotubes

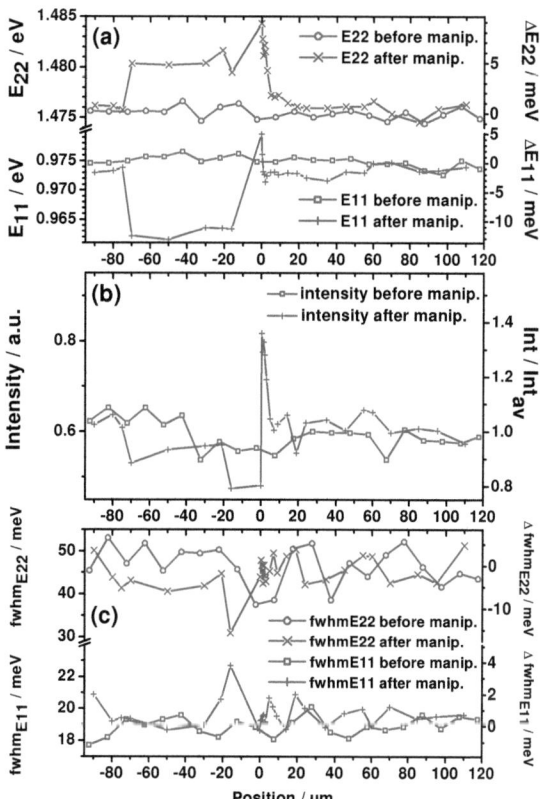

Figure 5.25: (a) E_{ii}, (b) intensity and (c) fwhm$_{Eii}$ ($i = 1, 2$) as a function of position before (gray) and after (red) manipulation of SWNT 2.

ΔE_{22} positive. We consider this an indication of week torsional strain. Note that this is not a result of uniaxial strain as this leads to an increase of E_{11} and a decrease of E_{22} as seen nicely on the left side, where the dirt particle caused a big shift of up to 27 and -28 meV for E_{11} and E_{22} due to uniaxial strain, respectively. Neither is it a 'tail' of the peak at Pos. 0 as this peak is strongly localized and shows an increase of E_{11}. Note also that the emission wavelength of this SWNT was close to the detection limit of the detector. Therefore, the decrease of E_{11} between Pos. 0.5-3 is possibly stronger than the data suggest. Between Pos. 3 and 10, the situation reverses and for Pos. 10-17, E_{11} is essentially back to its original value whereas ΔE_{22} is again positive before leveling off,

5 Spectroscopic Characterization of Individual Carbon Nanotubes

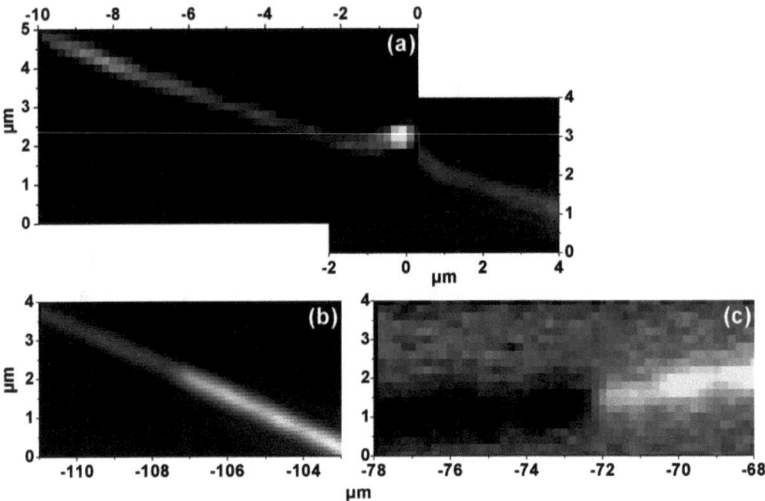

Figure 5.26: PL images of SWNT 1 and 2 after AFM manipulation. (a) PL image of SWNT 1 around the point of manipulation. Spectral integration was performed over a broad range to account for the large shift around Pos. 0. A localized burst of intensity can be seen around the point of manipulation. (b) PL image of SWNT 1 around Pos. -107 where the uniaxially strained part (on the right, bright) suddenly returns to the unstrained part (on the left, dim). The change in intensity is due to a Raman band superimposing the PL signal in the uniaxially strained (and therefore shifted) section. (c) Similar situation as in (b) for SWNT 2 around Pos. -72: Uniaxial strain (on the right, bright) is suddenly relieved (on the left, black). Here, the color code represents the shift of the center of mass of the PL emission signal.

too. SWNT 2 shows a similar but not so clear indication of torsion on its right side, between Pos. 0.5 and 2.5.

We now consider the strong increase of E_{11} (62 and 5 meV) and E_{22} (85 and 9 meV) around Pos. 0, accompanied by an increase of intensity by a factor of ∼7 and 1.4 for SWNT 1 and 2, respectively. The increased intensity at the point of manipulation is also nicely seen in the PL image of Fig. 5.26(a). An explanation that would fit qualitatively is that a short segment of the nanotubes around the point of manipulation is not touching the surface and is therefore behaving like a suspended SWNT. Quantitatively however, shifts do not match values listed in Tab. 5.2. It is also hard to imagine why the SWNTs

5.4 Spectroscopy and AFM Manipulation of Individual, Ultralong Carbon Nanotubes

should stick out of the surface after manipulation and SEM images (not shown) also do not indicate air-suspended sections. We therefore consider this scenario unlikely. We believe that the AFM tip inflicted irreversible damage on both SWNTs, including bond breaking. The extent of manipulation is bigger for SWNT 1, therefore the consequences are more pronounced: shifts in E_{ii} and intensity are larger, more abrupt and we see PL intermittency (blinking) at Pos. -0.5. A time trace of the PL is shown in the inset of Fig. 5.24(c). The blinking is very localized and found to become more pronounced by increasing laser excitation power (not shown). Note that up to now, we (and others, see Sec. 5.2 and Ref. [176]) have only witnessed blinking of micelle-wrapped SWNTs deposited on a surface below 25 K. Suspended and SWNTs with direct surface contact did not show PL intermittency even at low temperatures [10]. Matsuda *et al.* have seen blinking of micelle-wrapped SWNTs also at room temperature [179]. The results presented here now suggest that PL intermittency is not only of 'extrinsic' origin and related to a particular SWNT environment (e. g. surfactant coating as shown in Sec. 5.2), but can be linked to 'intrinsic' causes like defects, as well. Defects could be introduced by tip-sonication which is always needed for debundling. The reason for the increase of E_{ii} and intensity for both SWNTs is not understood at this stage. We see these results as a hint to investigate the connection between defects (kind of defects and concentration) on one side, and PL shifts and PL quantum yield on the other. Clearly, more research in this direction is needed.

In general, there was only one emission peak in the PL spectra of SWNT 1 and 2 as one would expect when individual SWNTs are probed. However, about 8 µm away from Pos. 0 of SWNT 1, we noticed a weak shoulder to the main peak on both sides of the nanotube. The positions of these shoulders are indicated by two red dots in Fig. 5.24(a). They are at almost identical distances to the point of manipulation. Their spectral position is close to that at Pos. 0. Thus, we believe that the existence of these shoulders is linked to the problem discussed in the previous paragraph. However, their origin is unclear at the moment.

We now turn to the discussion of uniaxial strain introduced by AFM manipulation. As already mentioned, SWNT 1 shows two sources of uniaxial strain, one due to a dirt particle being pushed over the tube and one due to the intentional manipulation with the AFM tip. The introduction of strain with the particle was prior to the AFM manipulation. The possible cutting of the SWNT at Pos. 0 may have led to a partial

5 Spectroscopic Characterization of Individual Carbon Nanotubes

relief of uniaxial strain for a position between -32 and 20. The uniaxial shifts due to the particle are 27 and -28 meV for E_{11} and E_{22}, respectively. For SWNT 2, ΔE_{11} = -13 meV and ΔE_{22} = +6 meV. In both cases, the family assignment is consistent with the signs of the measured shifts. Comparison of the two SWNTs shows that a large particle (or a blunt AFM tip) is more effective in uniaxially straining a carbon nanotube. Sharp tips tend to exert too much force on a small area and the tube ruptures if surface adhesion is strong, as in our case. Changes of E_{11} under small uniaxial strains (ϵ, not to be confused with the dielectric constant) are given by

$$\Delta E_{11} = sgn(2p+1)3|\gamma_0|(1+\nu)\epsilon \cos 3\theta \tag{5.3}$$

where $\gamma_0 \approx 2.7$ eV is the nearest-neighbor exchange integral, $\nu \approx 0.2$ Poisson's ratio, θ the chiral angle of the SWNT and $p = -1, 0, 1$ is the $(n-m)$ mod 3 family [201]. Using this equation the uniaxial strain is about 0.3% and 0.14% for SWNT 1 and SWNT 2, respectively.

In agreement with Duan et al., uniaxial strain has a much longer range than torsional strain. They report a length scale of up to 350 µm for uniaxial and ~ 12 µm for torsional strain [195]. We estimate for both SWNTs the range to be up to ~ 70 µm and ~ 5 µm for uniaxial and torsional strain, respectively. The length scale of uniaxial strain is easier to specify than of torsional strain, because the former shows an abrupt return to the unstrained value at Pos. -107 for SWNT 1 and at Pos. -72 and 58 for SWNT 2, in contrast to Duan et al., who report a continuous relaxation of uniaxial strain. PL images of the area around Pos. -107 of SWNT 1 and Pos. -72 of SWNT 2 in Fig. 5.26(b) and (c) show a sudden return to the unstrained tube. The color code in Fig. 5.26(b) represents the integrated area over both original and strained PL emission peak. The higher intensity and therefore brighter color, which can also be seen in Fig. 5.24(c), is due to a Raman signal superimposing the E_{11}/E_{22} electronic peak only in the uniaxially strained (and therefore shifted) area. In general, PL intensity is not affected by uniaxial strain, as seen for SWNT 2 in Fig. 5.25(b).

Regarding a dependency of the line widths on strain, we measure a small average increase of fwhm$_{E11}$ of 1-2 meV for both SWNT 1 and 2. However, the scatter of the data is too large for a definite assessment. Again, the broadening of the E_{11} emission signal between Pos. -26 and -106 for SWNT 1 is caused by a Raman band. For fwhm$_{E22}$ there seems to be no clear trend.

5.5 Individual SWNTs in Dispersion: A Method to Count Different Chiralities

Apart from the PL of individual, surface-bound (Sec. 5.4 and 5.2), air-suspended (Sec. 5.1, 5.2 and 5.3) and SWNTs deposited from dispersion (Sec. 5.2 and Ref. [111]), we were also able to observe individual emitters in diluted aqueous dispersions of nanotubes with our PL microscope [13]. This approach could be of interest from both an application-oriented and a fundamental point of view: The NIR emission of SWNTs is in a spectral range where biological tissue is almost transparent (the long wavelength reduces scattering). In addition, SWNTs show a much better photostability than currently used fluorescent markers. Disadvantages are the low quantum yield and a possible toxicity of nanotubes (or of surfactant or of residual catalyst particles, respectively). Weisman *et al.* have inserted and traced SWNTs in cells, fruit flies and rabbits via PL spectroscopy and imaging techniques on the individual tube level [202–204]. They observe no acute toxicity or adverse effects from low-level nanotube exposure.

From a fundamental standpoint, measuring individual SWNTs in a serial fashion opens the possibility to gain chirality-selective information about semiconducting nanotubes in an ensemble of SWNTs. This could include relative concentrations deduced by counting PL events in one chirality specific spectral range as opposed to another. When compared to ensemble PL spectra, the relative concentrations could lead to relative quantum yields of certain (n_1, n_2) species in dispersion.

Two possible set-ups [denoted (1) and (2)] for a measurement of individual SWNTs in dispersion are shown in Fig. 5.27. In set-up (1), a small drop of a diluted dispersion (typically between 1:5 and 1:50, determined empirically and diluted with D_2O/surfactant, with the same surfactant concentration as in the original dispersion, typically 1 wt%) is compressed between a sapphire wafer and a cover slip, resulting in a calculated film thickness of $\sim 5\,\mu m$. Set-up (2) uses a fused silica capillary (Polymicro TECHNOLOGIES) through which the dispersion can be pumped by a syringe (Hamilton) and a syringe pump (TSE systems, '540101'). The silica capillaries used have inner diameters between 2 and 15 µm and outer diameters of 360 µm, including an about 20- µm-thick protective polyimide coating. Along a section of a few millimeters in the middle of the tube, the coating is removed to create a 'window' for the measurement. The capillary is clamped on both sides of the Leica oil immersion objective which is used in set-up (1)

5 Spectroscopic Characterization of Individual Carbon Nanotubes

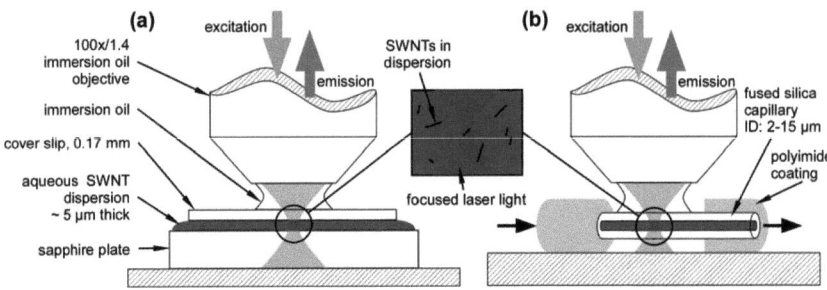

Figure 5.27: Set-ups for PL spectroscopy of individual SWNTs in dispersion. **(a)** Set-up (1) **(b)** Set-up (2). ID denotes the inner diameter (see text for details).

and (2) to focus the excitation light into the dispersion. A disadvantage of set-up (1) is the evaporation of D_2O which limits the measurement time. This is not a problem for set-up (2). In both set-ups, we believe to have frequently observed partially immobilized SWNTs due to their adsorption either on the silica tube in set-up (2) or on the sapphire wafer/cover slip in set-up (1). The PL signal of an individual SWNT in this case varied in intensity (and sometimes vanished and reemerged) but the same tube (presumably) remained in the detection area for several seconds. Observations by Weisman et al. using a 2D InGaAs imaging detector confirm this view [205]. They also state that only nanotubes that are stationary during the acquisition time can be clearly imaged. This might be the reason why in our experiments, the strongest signals were acquired with the focus close to one of the interface layers dispersion/silica or dispersion/sapphire. The diffusion in the medium could be too fast at room temperature for typical used acquisition times (between 0.1 and 1 s). Weisman et al. measured a diffusion constant of $D = 4\,\mu m^2/s$ [205] but they do not state if the tube is adsorbed on a surface or not.

To avoid repeated measurements of the same tube, the position of the sample could be changed by the piezo stage. However, this would require either a precise adjustment of the sample or corrections in the z-direction of the focus. Figure 5.28 shows a characteristic multi-spectra file of a HiPco/SDBS dispersion acquired with set-up (2), where the sample position along the silica tube was changed repeatedly. The complete file consists of 6000 successively recorded spectra at an excitation wavelength of 730 nm and an acquisition time of 1 s per spectrum. For the sake of clarity, only the first 600 are depicted.

5.5 Individual SWNTs in Dispersion: A Method to Count Different Chiralities

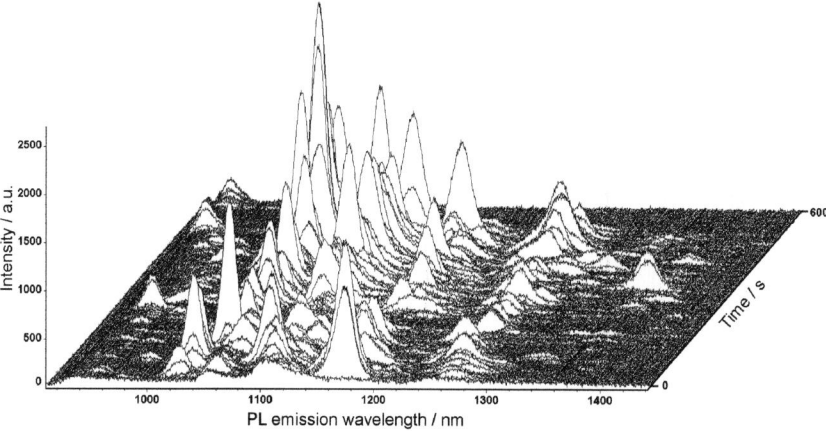

Figure 5.28: Multi-spectra file showing the first 600 spectra out of 6000, recorded of a HiPco/SDBS dispersion at an excitation wavelength of 730 nm. Each spectrum was acquired for 1 s. To ensure proper averaging, the sample position was manually altered repeatedly during the measurement.

The spectra were then analyzed in two ways with the 'Origin' software (OriginLab Corporation). First, we checked if the number of spectra was representative for the ensemble. We therefore applied the ergodic hypothesis which states that the average of a process parameter over time and the average over the statistical ensemble become equivalent if time goes to infinity. However, adding up 6000 spectra resulted in a very noisy spectrum which was mainly due to the read-out noise from the detector. In contrast to statistical noise, read-out noise is often equivalent from one spectrum to the other and thus does not cancel when summed up. We reduced the noise by introducing a threshold of 50 counts. In every spectrum, all signals below the threshold were set to zero and all signals above were reduced by 50 counts. After the summation, the red spectrum of Fig. 5.29(a) was obtained. To further reduce the noise, the red spectrum was smoothed (green curve). When compared to the ensemble spectrum (black curve, normalized to the (9, 4) SWNT), the overall agreement is already fairly good. The (7, 5) tube is somewhat too pronounced whereas the intensity of the (8, 7) SWNT is slightly reduced compared to the ensemble spectrum. Performing the same procedure with only

5 Spectroscopic Characterization of Individual Carbon Nanotubes

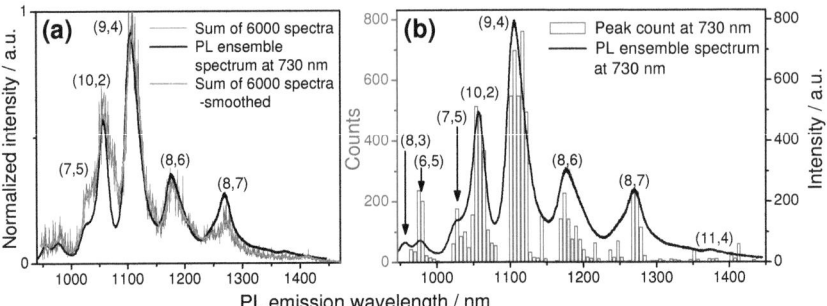

Figure 5.29: Results from the analysis of 6000 successive PL spectra partially shown in Fig. 5.28. **(a)** Test of the ergodic hypothesis by summing all 6000 spectra (red curve) and comparing them to a scaled ensemble spectrum (black curve) after smoothing (green curve). **(b)** Histogram of counted PL events, again referenced to a PL ensemble spectrum (see text for details).

600 spectra showed a much larger mismatch of relative intensities. Thus, we believe that the 6000 spectra represented a good statistical basis for the 'counting analysis'.

The counting analysis produced a histogram as shown in Fig. 5.29(b) by counting the peaks whose maxima fell into a given interval. The software code defines a peak as a user-defined change of intensity within a user-defined spectral width. In Fig. 5.29(b), a peak was counted as a peak, when the intensity dropped by 50 counts ±15 nm from its maximum position. Within a spectral region fulfilling this condition, the routine takes the position with the larges intensity. The amount of counted nanotubes that resulted from the choice of these two parameters ('intensity drop' and 'spectral width') could be quite different, but we believe and checked that our choice was quite reasonable[4]. The size of the interval for the histogram was chosen to be ~5 nm in Fig. 5.29(b) and the ensemble spectrum at the same excitation wavelength, normalized to the (9,4) tube, is shown as well. The number of counts are now a measure for the relative abundances, whereas the spectral intensity of the ensemble spectrum shows the spectral weight of different chiralities. The SWNTs with chiral indices (10, 2), (9, 4), (8, 6) and (8, 7) show λ_{22} values of 736, 726, 720 and 732 nm, respectively, and are thus closest to the excitation

---
[4] A conversion of all spectra to an equidistant energy scale instead of a wavelength scale or a more sophisticated peak-searching tool might further improve results.

5.5 Individual SWNTs in Dispersion: A Method to Count Different Chiralities

wavelength of 730 nm. The counts of the $(7,5), (10,2)$ and $(8,7)$ tube agree quite well with the intensities of the ensemble. Interestingly, the $(6,5)$ tube shows a much larger abundance compared to its spectral weight. The opposite holds for the $(8,6)$ SWNT. In total, about 8000 PL events were counted which corresponds to about 1.3 detected nanotubes per second. To yield the relative concentration of a chiral species, one could add the number of counts that fall into the emission region of that tube. However, care has to be taken when doing so, because different chiralities with different E_{22} excitation energies but almost equal emission energies tend to overlap. For instance, the $(7,5)$ tube that is responsible for the shoulder of the $(10,2)$ SWNT has its absorption maximum at 880 nm. Nevertheless, it is a prominent feature even at an excitation wavelength of 730 nm and its counts should not be confused with those of the $(10,2)$ tube. The reason why it absorbs so strongly could be due to a phonon-assisted transition (exciton-phonon complex, see Sec. 2.4.4).

After the determination of the relative abundances, the relative quantum yields are of interest, as well. If the PL spectra of two ensembles of individualized and (n_1, n_2)-monodispersed nanotubes, denoted SWNT A and SWNT B were measured with the same set-up and the same excitation power at an excitation wavelength λ_{exc}, the following would hold for the ratio of their spectrally integrated intensities I_i with $i = A, B$:

$$\frac{I_A}{I_B} = \frac{\int PL_{A,(\lambda)} \cdot IRF_{(\lambda)}^{-1} d\lambda}{\int PL_{B,(\lambda)} \cdot IRF_{(\lambda)}^{-1} d\lambda} = \frac{\sigma_{(\lambda_{exc}, A)}}{\sigma_{(\lambda_{exc}, B)}} \cdot \frac{\eta_A}{\eta_B} \cdot \frac{c_A}{c_B} \qquad (5.4)$$

where $PL_{i,(\lambda)}$ denotes the PL spectrum, $IRF_{(\lambda)}$ the **I**nstrument **R**esponse **F**unction, $\sigma_{(\lambda_{exc}, i)}$ the absorption cross section, η_i the PL quantum yield and c_i the concentration of tube i. The ratio c_A/c_B could be calculated from the counting analysis, I_A/I_B from an ensemble spectrum (although the superposition of chiralities would not allow for an unambiguous fitting) and $\sigma_{(\lambda_{exc}, A)}/\sigma_{(\lambda_{exc}, B)}$ from absorption spectra (same problem of overlapping absorption peaks).

Note that the absorption cross section of nanotubes critically depends on the excitation wavelength. For instance, $\sigma_{(730\,nm)}$ of the $(6,5)$ tube which was excited off-resonance ($\lambda_{22} = 566$ nm) in the ensemble spectrum of Fig. 5.29 can be expected to be much lower than that of the tubes that are close to resonance [$(10,2), (9,4), (8,6)$ and $(8,7)$]. Therefore, only SWNTs with their absorption maximum close to the excitation wavelength should be compared. The $(8,3)$ SWNT was too close to the left edge of the spectrum to be detected by the peak-finding tool and the $(11,4)$ tube was already hardly present in the ensemble.

The measurements presented in this section are to be regarded as still preliminary, as the amount of experimental data is still limited and their statistical treatment is difficult and computationally challenging. The confidence in this method could be increased further if it was first tested on a system like dispersed quantum dots of known concentration and quantum yield. However, this section shows that the basic principle of counting SWNTs by the successive acquisition of PL spectra works and that first results are promising. One could envision an automatic measurement of histograms for preselected excitation wavelengths to account for different chiralities or the use of dispersions enriched with a single chirality. Dispersions of only a few, spectrally well separated (n_1, n_2) species in toluene are now available and could prove to be very valuable in this context, as well. We are also working on the implementation of a 2D InGaAs imaging detector into our microscope which could show whether the tubes in our measurements are adsorbed to a surface or not.

6 Conclusion and Outlook

In summary, the thermal CVD synthesis of nanotubes and their spectroscopic (with the focus on PL microscopy) and AFM characterization and manipulation on an individual tube level were presented in this thesis. For the nanotube synthesis, a thermal CVD reactor employing an array of different carbon feedstock substances was designed and built. For PL(E) measurements on individual nanotubes or ensembles thereof, an automated PL laser microscope was built with the capability to perform PL imaging and PLE spectroscopy in the excitation wavelength range of $\sim$ 600-1000 nm and emission range of $\sim$ 800-1600 nm.

The growth of bulk amounts of VANTAs was described in Sec. 4.1, where sputtered Al_2O_3/Fe bilayers with a characteristic thickness of 10/1 nm were exposed to gas mixtures of ethylene, argon, hydrogen and water vapor at temperatures around 750 °C. By optimizing parameters like gas/vapor composition, flow rate, heat-up rate and temperature, maximum forest heights of 1.3 mm in 90 minutes were achieved. The VANTAs consisted mostly of MWNTs with diameters between 3-4 nm. However, about 8 % were SWNTs with diameters below 2 nm. Those SWNTs were partially suspended on top of the forests and were particularly suited for PL and Raman measurements.

Besides VANTAs, other growth morphologies like individual nanotubes air-suspended over cracks of the catalyst material or resting on a wafer surface (SiO_2 or sapphire) and aligned with the gas flow were realized, as well (Sec. 4.2). These methods use ethanol, carbon monoxide or methane as carbon feedstock gases and a catalyst prepared from a dispersion of iron nitrate, molybdenyl(VI)acetylacetonate and alumina in methanol or isopropanol. Samples from the above preparations were used for spectroscopic studies summarized below.

In Sec. 5.1, µ-PLE spectroscopy of individual SWNTs, air-suspended on top of VANTAs was described. Observed optical transition energies E_{11} and E_{22} were compared to those of nanotubes in aqueous dispersions. A blueshift of 40-56 meV for E_{11} and

6 Conclusion and Outlook

24-48 meV for E_{22} was found and assigned to 19 different chiralities. A significant correlation between tube diameter or chirality and energy shift was not observed. The upshift in energy was explained by different dielectric screening of excitons in a water-surfactant environment as opposed to air or vacuum. Measurements of SWNTs (on top of VANTAs) immersed in paraffin oil and 1-methylnaphthalene showed similar transition energies as surfactant-coated nanotubes in aqueous dispersions which implies a similar dielectric screening for both nanotube surroundings. Mean shifts (referenced to SWNTs in D_2O/1 wt% SDBS) were -1 meV and -14 meV for E_{11} and E_{22} for 1-methylnaphthalene and -13 meV and -27 meV for paraffin oil, respectively.

Low temperature µ-PLE spectroscopy on (i) individual SWNTs air-suspended on top of VANTAs and between catalyst cracks from carbon monoxide CVD; (ii) HiPco nanotubes, deposited from dispersion and (iii) surface-bound SWNTs on sapphire from carbon monoxide CVD was described in Sec. 5.2. PL intermittency and spectral diffusion was observed for surfactant-coated nanotubes at temperatures below 25 K, whereas surfactant-free SWNTs showed stable PL emission. This implies that blinking and spectral diffusion can be caused by a surfactant coating. Small ensembles of contacting SWNTs exhibit much more complex PLE spectra (including enhanced sidebands and shifts) than those from isolated emitters. PLE maxima of deposited surfactant-coated SWNTs at ambient temperature are close to the ensemble values of SWNTs in the initial dispersion whereas as-grown nanotubes directly contacting the surface show blueshifts of 15-30 meV and 10-25 meV for E_{11} and E_{22} compared to nanotubes in aqueous dispersion, respectively. These correlations were also applicable at cryogenic temperatures when taking into account small additional blueshifts of a few meV due to the temperature decrease. We were thus able to assign (n_1, n_2) values to most SWNTs of samples (i)-(iii) at all temperatures. On average, PL intensities (at all temperatures) and therefore PL quantum yields were found to scale as $(i) : (ii) : (iii) = 100 : 50 : 5$.

The first direct observation of weakly optically allowed excitonic states below E_{11} was described in Sec. 5.3. µ-PLE spectra of individual, air-suspended SWNTs revealed emission satellites, redshifted by 38-45 and 100-130 meV and $\sim$ 30-300 times less intense compared to the main E_{11} emission signals. Only one emission satellite 90-143 meV below E_{11} for different (n_1, n_2) species could be resolved in dispersed, aqueous SWNT ensembles, with $\sim$ 10 times less intensity, but recent measurements on SWNTs dispersed in toluene have confirmed the first emission satellites, as well.

In Sec. 5.4, individual, as-grown SWNTs from carbon monoxide, ethanol and methane CVD were characterized by μ-PLE spectroscopy, μ-Raman spectroscopy and AFM along their z-axis, i.e. along their length. Most CVD-grown nanotubes of all studied samples had diameters too large for PLE spectroscopy and there was also a considerable fraction of long CO-CVD tubes that presumably grew as bundles. The possibility of finding peculiar configurations of nanotubes like loops with SEM, AFM and μ-Raman spectroscopy, as well as the dependency of the emission intensity on the relative polarization of the excitation light and the nanotube z-axis was demonstrated. Very long sections of SWNTs from ethanol and methane CVD were found to show a remarkable integrity of (n_1, n_2) chirality over lengths between 50-600 μm. We observed only one SWNT which presumably changed its chiral indices during growth. Nanotubes grown over lithographically produced trenches and markers showed blueshifts between 7-27 meV and 5-24 meV for E_{11} and E_{22} when going from surface-bound to air-suspended sections, respectively, and PL intensity (PL quantum yield) increased by a factor of 10-20, in agreement with measurements in Sec. 5.2. The former effect is again due to changes in the dielectric surroundings of the SWNTs whereas the latter is rationalized in terms of an effective quenching of excitons for nanotubes in contact with a dielectric surface. Line width data scattered too much to derive a dependence of fwhm$_{Eii}$ ($ii = 1, 2$) on the dielectric constant ϵ_r, but we see indications of a decrease of fwhm$_{E11}$ for increasing tube diameter and an increase of fwhm$_{E22}$ with increasing E_{22}.

AFM manipulation of two unsuspended sections of SWNTs showed characteristic shifts of uniaxial and torsional strain but also localized severe defects like tube fracture. Such defects led to PL intermittency at room temperature of one SWNT, to large shifts in E_{ii} and, surprisingly, to an increase of PL intensity. None of these effects had been reported so far. Residual uniaxial and torsional strain could be detected up to 70 μm and 5 μm away from the fractured site, respectively. The uniaxially strained sections of the nanotubes returned to the unstrained values on a length scale shorter than the resolution of our PLE microscope (~ 400 nm) as opposed to a gradual relaxation.

In the last section, a new method for counting individual chiralities of SWNTs in dispersion was presented and preliminary results were discussed. The method could help to determine the relative concentrations of different chiralities of semiconducting nanotubes and eventually, when combined with chirality specific absorption cross sections, relative quantum yields.

6 Conclusion and Outlook

In the near future, we plan to upgrade our µ-PLE set-up with a 2D InGaAs camera which would allow much faster imaging capabilities compared to the presently used method of raster-scanning an area pixel by pixel. This would greatly increase the speed at which further samples of ultralong, aligned SWNTs from Li *et al.* could be investigated. Nanotubes show unusual and complex changes in the PL properties after severe AFM manipulation. These effects and a possible influence of the substrate require further experimental investigations. One could also envision the use of tools other than AFM for manipulation [e. g. cutting with an electron beam or by fast ion bombardment (FIB)]. In the near future of PLE spectroscopy of carbon nanotubes, aqueous SWNT dispersions will probably be replaced by toluene dispersions where the surfactant is a fluorine-based organic polymer. Such dispersions are superior to currently used water-surfactant dispersions in many respects e. g. quantum yield and selectivity. Chirality-selected dispersions of spectrally well-separated nanotubes in toluene could be used as model systems in the 'counting analysis'.

Bibliography

[1] Iijima, S. *Nature* **1991**, *354*, 56.

[2] Bachilo, S. M.; Strano, M. S.; Kittrell, C.; Hauge, R. H.; Smalley, R. E.; Weisman, R. B. *Science* **2002**, *298*, 2361.

[3] Krishnan, A.; Dujardin, E.; Ebbesen, T. W.; Yianilos, P. N.; Treacy, M. M. J. *Phys. Rev. B* **1998**, *58*, 14013.

[4] Collins, P. G.; Avouris, P. *Scientific American* **2000**, *283*, 62.

[5] Baughman, R. H.; Zakhidov, A. A.; de Heer, W. A. *Science* **2002**, *297*, 787.

[6] Jorio, A., Dresselhaus, M. S., Dresselhaus, G., Eds. *Carbon Nanotubes-Advanced Topics in the Synthesis, Structure, Properties and Applications*; Topics in Applied Physics 111; Springer-Verlag, 2008.

[7] Wang, F.; Dukovic, G.; Brus, L. E.; Heinz, T. F. *Science* **2005**, *308*, 838.

[8] Maultzsch, J.; Pomraenke, R.; Reich, S.; Chang, E.; Prezzi, D.; Ruini, A.; Molinari, E.; Strano, M. S.; Thomsen, C.; Lienau, C. *Phys. Rev. B* **2005**, *72*, 241402.

[9] Kiowski, O.; Lebedkin, S.; Hennrich, F.; Malik, S.; Rösner, H.; Arnold, K.; Sürgers, C.; Kappes, M. M. *Phys. Rev. B* **2007**, *75*, 075421.

[10] Kiowski, O.; Lebedkin, S.; Hennrich, F.; Kappes, M. M. *Phys. Rev. B* **2007**, *76*, 075422.

[11] Kiowski, O.; Arnold, K.; Lebedkin, S.; Hennrich, F.; Kappes, M. M. *Phys. Rev. Lett.* **2007**, *99*, 237402.

Bibliography

[12] Kiowski, O.; Jester, S. S.; Lebedkin, S.; Yin, Z.; Li, Y.; Kappes, M. M. *Photoluminescence Spectral Imaging of Ultra-Long Single-Walled Carbon Nanotubes on Silicon: Micromanipulation Induced Strain, Torsion, Rupture and the Determination of Handedness*; submitted to Nano Letters.

[13] Kiowski, O.; Arnold, K.; Lebedkin, S.; Hennrich, F.; Kappes, M. M. In *Electronic Properties of Novel Nanostructures*; Kuzmany, H., Fink, J., Mehring, M., Roth, S., Eds.; AIP, 2005; Vol. 786; p 139.

[14] Reich, S.; Thomsen, C.; Maultzsch, J. *Carbon Nanotubes: Basic Concepts and Physical Properties*; Wiley-VCH, 2004.

[15] Zheng, L. X.; O'Connell, M. J.; Doorn, S. K.; Liao, X. Z.; Zhao, Y. H.; Akhadov, E. A.; Hoffbauer, M. A.; Roop, B. J.; Jia, Q. X.; Dye, R. C.; Peterson, D. E.; Huang, S. M.; Liu, J.; Zhu, Y. T. *Nat. Mater.* **2004**, *3*, 673.

[16] Oberlin, A.; Endo, M.; Koyama, T. *J. Cryst. Growth* **1976**, *32*, 335–349.

[17] Radushkevich, L. V.; Lukyanovich, V. M. *Zurn. Fisic. Chim.* **1952**, *26*, 88–95.

[18] Iijima, S.; Ichihashi, T. *Nature* **1993**, *363*, 603.

[19] Bethune, D. S.; Kiang, C. H.; Devries, M. S.; Gorman, G.; Savoy, R.; Vazquez, J.; Beyers, R. *Nature* **1993**, *363*, 605.

[20] Monthioux, M.; Kuznetsov, V. L. *Carbon* **2006**, *44*, 1621.

[21] Samsonidze, G. G.; Saito, R.; Jorio, A.; Pimenta, M. A.; Souza, A. G.; Gruneis, A.; Dresselhaus, G.; Dresselhaus, M. S. *Journal of Nanoscience and Nanotechnology* **2003**, *3*, 431.

[22] Saito, R.; Dresselhaus, G.; Dresselhaus, M. S. *Physical Properties of Carbon Nanotubes*; Imperial College Press, 1998.

[23] Mintmire, J. W.; White, C. T. *Phys. Rev. Lett.* **1998**, *81*, 2506.

[24] Kataura, H.; Kumazawa, Y.; Maniwa, Y.; Umezu, I.; Suzuki, S.; Ohtsuka, Y.; Achiba, Y. *Synthetic Met.* **1999**, *103*, 2555.

[25] Saito, R.; Dresselhaus, G.; Dresselhaus, M. S. *Phys. Rev. B* **2000**, *61*, 2981.

[26] Holden, J. M.; Ping, Z.; Bi, X. X.; Eklund, P. C.; Bandow, S. J.; Jishi, R. A.; Daschowdhury, K.; Dresselhaus, G.; Dresselhaus, M. S. *Chem. Phys. Lett.* **1994**, *220*, 186.

[27] Rao, A. M.; Richter, E.; Bandow, S.; Chase, B.; Eklund, P. C.; Williams, K. A.; Fang, S.; Subbaswamy, K. R.; Menon, M.; Thess, A.; Smalley, R. E.; Dresselhaus, G.; Dresselhaus, M. S. *Science* **1997**, *275*, 187.

[28] Journet, C.; Bernier, P. *Appl. Phys. A-Mater.* **1998**, *67*, 1.

[29] Girifalco, L. A.; Hodak, M.; Lee, R. S. *Phys. Rev. B* **2000**, *62*, 13104.

[30] Arnold, K. Experimentelle Untersuchungen elektronischer Eigenschaften von Kohlenstoffnanoröhren mit Hilfe der Photolumineszenzspektroskopie. Diploma Thesis, Institut für Physikalische Chemie, Universität Karlsruhe (TH), 2002.

[31] Bachilo, S. M.; Balzano, L.; Herrera, J. E.; Pompeo, F.; Resasco, D. E.; Weisman, R. B. *J. Am. Chem. Soc.* **2003**, *125*, 11186.

[32] Arnold, M. S.; Stupp, S. I.; Hersam, M. C. *Nano Lett.* **2005**, *5*, 713.

[33] Arnold, M. S.; Green, A. A.; Hulvat, J. F.; Stupp, S. I.; Hersam, M. C. *Nat. Nanotechnol.* **2006**, *1*, 60.

[34] Nish, A.; Hwang, J. Y.; Doig, J.; Nicholas, R. J. *Nat. Nanotechnol.* **2007**, *2*, 640.

[35] Lebedkin, S.; Hennrich, F.; Kiowski, O.; Kappes, M. M. *Phys. Rev. B* **2008**, *77*, 165429.

[36] Sfeir, M. Y.; Wang, F.; Huang, L. M.; Chuang, C. C.; Hone, J.; O'Brien, S. P.; Heinz, T. F.; Brus, L. E. *Science* **2004**, *306*, 1540.

[37] Sfeir, M. Y.; Beetz, T.; Wang, F.; Huang, L. M.; Huang, X. M. H.; Huang, M. Y.; Hone, J.; O'Brien, S.; Misewich, J. A.; Heinz, T. F.; Wu, L. J.; Zhu, Y. M.; Brus, L. E. *Science* **2006**, *312*, 554.

[38] Wang, F.; Sfeir, M. Y.; Huang, L. M.; Huang, X. M. H.; Wu, Y.; Kim, J. H.; Hone, J.; O'Brien, S.; Brus, L. E.; Heinz, T. F. *Phys. Rev. Lett.* **2006**, *96*, 167401.

[39] Qiu, X. H.; Freitag, M.; Perebeinos, V.; Avouris, P. *Nano Lett.* **2005**, *5*, 749.

[40] Misewich, J. A.; Martel, R.; Avouris, P.; Tsang, J. C.; Heinze, S.; Tersoff, J. *Science* **2003**, *300*, 783.

[41] Freitag, M.; Chen, J.; Tersoff, J.; Tsang, J. C.; Fu, Q.; Liu, J.; Avouris, P. *Phys. Rev. Lett.* **2004**, *93*, 076803.

[42] Bozovic, I.; Bozovic, N.; Damnjanovic, M. *Phys. Rev. B* **2000**, *62*, 6971.

[43] Damnjanovic, M.; Vukovic, T.; Milosevic, I. *Solid State Communications* **2000**, *116*, 265.

[44] Vukovic, T.; Milosevic, I.; Damnjanovic, M. *Phys. Rev. B* **2002**, *65*, 045418.

[45] Gruneis, A.; Saito, R.; Samsonidze, G. G.; Kimura, T.; Pimenta, M. A.; Jorio, A.; Souza, A. G.; Dresselhaus, G.; Dresselhaus, M. S. *Phys. Rev. B* **2003**, *67*, 165402.

[46] Ajiki, H.; Ando, T. *Japanese Journal of Applied Physics Part 2-Letters* **1994**, *34*, 107.

[47] Benedict, L. X.; Louie, S. G.; Cohen, M. L. *Phys. Rev. B* **1995**, *52*, 8541.

[48] Tasaki, S.; Maekawa, K.; Yamabe, T. *Phys. Rev. B* **1998**, *57*, 9301.

[49] Kittel, C. *Einführung in die Festkörperphysik*; Oldenbourg Verlag, 2002.

[50] Murakami, Y.; Chiashi, S.; Einarsson, E.; Maruyama, S. *Phys. Rev. B* **2005**, *71*, 085403.

[51] Miyauchi, Y.; Oba, M.; Maruyama, S. *Phys. Rev. B* **2006**, *74*, 205440.

[52] Miyauchi, Y.; Maruyama, S. *Phys. Rev. B* **2006**, *74*, 035415.

[53] Hertel, T. Optical properties of carbon nanotubes. presentation at the Summer School on Nanotubes, Cargese, Corsica, 3-15 July 2006 (unpublished).

[54] Barros, E. B.; Capaz, R. B.; Jorio, A.; Samsonidze, G. G.; Souza, A. G.; Ismail-Beigi, S.; Spataru, C. D.; Louie, S. G.; Dresselhaus, G.; Dresselhaus, M. S. *Phys. Rev. B* **2006**, *73*, 241406.

[55] Barros, E. B.; Jorio, A.; Samsonidze, G. G.; Capaz, R. B.; Souza, A. G.; Mendes, J.; Dresselhaus, G.; Dresselhaus, M. S. *Physics Reports-Review Section of Physics Letters* **2006**, *431*, 261.

[56] Jiang, J.; Saito, R.; Samsonidze, G. G.; Jorio, A.; Chou, S. G.; Dresselhaus, G.; Dresselhaus, M. S. *Phys. Rev. B* **2007**, *75*, 035407.

[57] Dresselhaus, M. S.; Dresselhaus, G.; Saito, R.; Jorio, A. *Annual Review of Physical Chemistry* **2007**, *58*, 719.

[58] Lauret, J. S.; Voisin, C.; Cassabois, G.; Delalande, C.; Roussignol, P.; Jost, O.; Capes, L. *Phys. Rev. Lett.* **2003**, *90*, 057404.

[59] Hagen, A.; Steiner, M.; Raschke, M. B.; Lienau, C.; Hertel, T.; Qian, H. H.; Meixner, A. J.; Hartschuh, A. *Phys. Rev. Lett.* **2005**, *95*, 197401.

[60] Tan, P. H.; Rozhin, A. G.; Hasan, T.; Hu, P.; Scardaci, V.; Milne, W. I.; Ferrari, A. C. *Phys. Rev. Lett.* **2007**, *99*, 137402.

[61] Qian, H. H.; Georgi, C.; Anderson, N.; Green, A. A.; Hersam, M. C.; Novotny, L.; Hartschuh, A. *Nano Lett.* **2008**, *8*, 1363.

[62] Pichler, T.; Knupfer, M.; Golden, M. S.; Fink, J.; Rinzler, A.; Smalley, R. E. *Phys. Rev. Lett.* **1998**, *80*, 4729.

[63] Itkis, M. E.; Niyogi, S.; Meng, M. E.; Hamon, M. A.; Hu, H.; Haddon, R. C. *Nano Lett.* **2002**, *2*, 155.

[64] Kane, C. L.; Mele, E. J. *Phys. Rev. Lett.* **2003**, *90*, 207401.

[65] O'Connell, M. J.; Bachilo, S. M.; Huffman, C. B.; Moore, V. C.; Strano, M. S.; Haroz, E. H.; Rialon, K. L.; Boul, P. J.; Noon, W. H.; Kittrell, C.; Ma, J. P.; Hauge, R. H.; Weisman, R. B.; Smalley, R. E. *Science* **2002**, *297*, 593.

[66] Ando, T. *J. Phys. Soc. Jpn.* **1997**, *66*, 1066.

[67] Mott, N. F. *Trans. Faraday Soc.* **1938**, *34*, 500–506.

[68] Wannier, G. H. *Physical Review* **1937**, *52*, 191.

Bibliography

[69] Frenkel, J. *Physical Review* **1931**, *37*, 1276.

[70] Spataru, C. D.; Ismail-Beigi, S.; Benedict, L. X.; Louie, S. G. *Phys. Rev. Lett.* **2004**, *92*, 077402.

[71] Spataru, C. D.; Ismail-Beigi, S.; Benedict, L. X.; Louie, S. G. *Appl. Phys. A-Mater.* **2004**, *78*, 1129.

[72] Rohlfing, M.; Louie, S. G. *Phys. Rev. B* **2000**, *62*, 4927.

[73] Spataru, C. D.; Ismail-Beigi, S.; Capaz, R. B.; Louie, S. G. *Phys. Rev. Lett.* **2005**, *95*, 247402.

[74] Zhao, H. B.; Mazumdar, S. *Phys. Rev. Lett.* **2004**, *93*, 157402.

[75] Perebeinos, V.; Tersoff, J.; Avouris, P. *Nano Lett.* **2005**, *5*, 2495.

[76] Zhao, H.; Mazumdar, S.; Sheng, C. X.; Tong, M.; Vardeny, Z. V. *Phys. Rev. B* **2006**, *73*, 075403.

[77] Tretiak, S. *Nano Lett.* **2007**, *7*, 2201.

[78] Lebedkin, S.; Arnold, K.; Hennrich, F.; Krupke, R.; Renker, B.; Kappes, M. M. *New J. Phys.* **2003**, *5*, 140.

[79] Wang, F.; Dukovic, G.; Brus, L. E.; Heinz, T. F. *Phys. Rev. Lett.* **2004**, *92*, 177401.

[80] Jones, M.; Engtrakul, C.; Metzger, W. K.; Ellingson, R. J.; Nozik, A. J.; Heben, M. J.; Rumbles, G. *Phys. Rev. B* **2005**, *71*, 115426.

[81] Lefebvre, J.; Austing, D. G.; Bond, J.; Finnie, P. *Nano Lett.* **2006**, *6*, 1603.

[82] Chou, S. G.; Plentz, F.; Jiang, J.; Saito, R.; Nezich, D.; Ribeiro, H. B.; Jorio, A.; Pimenta, M. A.; Samsonidze, G. G.; Santos, A. P.; Zheng, M.; Onoa, G. B.; Semke, E. D.; Dresselhaus, G.; Dresselhaus, M. S. *Phys. Rev. Lett.* **2005**, *94*, 127402.

[83] Plentz, F.; Ribeiro, H. B.; Jorio, A.; Strano, M. S.; Pimenta, M. A. *Phys. Rev. Lett.* **2005**, *95*, 247401.

[84] Htoon, H.; O'Connell, M. J.; Doorn, S. K.; Klimov, V. I. *Phys. Rev. Lett.* **2005**, *94*, 127403.

[85] Perebeinos, V.; Tersoff, J.; Avouris, P. *Phys. Rev. Lett.* **2005**, *94*, 027402.

[86] Saito, R.; Gruneis, A.; Samsonidze, G. G.; Brar, V. W.; Dresselhaus, G.; Dresselhaus, M. S.; Jorio, A.; Cancado, L. G.; Fantini, C.; Pimenta, M. A.; Souza, A. G. *New J. Phys.* **2003**, *5*, 157.1–157.15.

[87] Meyer, J. C.; Paillet, M.; Michel, T.; Moreac, A.; Neumann, A.; Duesberg, G. S.; Roth, S.; Sauvajol, J. L. *Phys. Rev. Lett.* **2005**, *95*, 217401.

[88] Jorio, A.; Saito, R.; Hafner, J. H.; Lieber, C. M.; Hunter, M.; McClure, T.; Dresselhaus, G.; Dresselhaus, M. S. *Phys. Rev. Lett.* **2001**, *86*, 1118.

[89] Milnera, M.; Kurti, J.; Hulman, M.; Kuzmany, H. *Phys. Rev. Lett.* **2000**, *84*, 1324.

[90] Fantini, C.; Jorio, A.; Souza, M.; Strano, M. S.; Dresselhaus, M. S.; Pimenta, M. A. *Phys. Rev. Lett.* **2004**, *93*, 147406.

[91] Jorio, A.; Fantini, C.; Pimenta, M. A.; Capaz, R. B.; Samsonidze, G. G.; Dresselhaus, G.; Dresselhaus, M. S.; Jiang, J.; Kobayashi, N.; Gruneis, A.; Saito, R. *Phys. Rev. B* **2005**, *71*, 075401.

[92] Maultzsch, J.; Reich, S.; Thomsen, C.; Webster, S.; Czerw, R.; Carroll, D. L.; Vieira, S. M. C.; Birkett, P. R.; Rego, C. A. *Appl. Phys. Lett.* **2002**, *81*, 2647.

[93] Brown, S. D. M.; Jorio, A.; Dresselhaus, M. S.; Dresselhaus, G. *Phys. Rev. B* **2001**, *64*, 073403.

[94] Saito, R.; Jorio, A.; Hafner, J. H.; Lieber, C. M.; Hunter, M.; McClure, T.; Dresselhaus, G.; Dresselhaus, M. S. *Phys. Rev. B* **2001**, *6408*, 085312.

[95] Dresselhaus, M. S.; Dresselhaus, G.; Saito, R.; Jorio, A. *Physics Reports-Review Section of Physics Letters* **2005**, *409*, 47.

[96] Jorio, A.; Souza, A. G.; Dresselhaus, G.; Dresselhaus, M. S.; Swan, A. K.; Unlu, M. S.; Goldberg, B. B.; Pimenta, M. A.; Hafner, J. H.; Lieber, C. M.; Saito, R. *Phys. Rev. B* **2002**, *65*, 155412.

[97] Ferrari, A. C.; Meyer, J. C.; Scardaci, V.; Casiraghi, C.; Lazzeri, M.; Mauri, F.; Piscanec, S.; Jiang, D.; Novoselov, K. S.; Roth, S.; Geim, A. K. *Phys. Rev. Lett.* **2006**, *97*, 187401.

[98] Gupta, A.; Chen, G.; Joshi, P.; Tadigadapa, S.; Eklund, P. C. *Nano Lett.* **2006**, *6*, 2667.

[99] Souza, A. G.; Jorio, A.; Swan, A. K.; Unlu, M. S.; Goldberg, B. B.; Saito, R.; Hafner, J. H.; Lieber, C. M.; Pimenta, M. A.; Dresselhaus, G.; Dresselhaus, M. S. *Phys. Rev. B* **2002**, *65*, 085417.

[100] Weisman, R. B.; Bachilo, S. M. *Nano Lett.* **2003**, *3*, 1235.

[101] Strano, M. S.; Doorn, S. K.; Haroz, E. H.; Kittrell, C.; Hauge, R. H.; Smalley, R. E. *Nano Lett.* **2003**, *3*, 1091.

[102] Telg, H.; Maultzsch, J.; Reich, S.; Hennrich, F.; Thomsen, C. *Phys. Rev. Lett.* **2004**, *93*, 177401.

[103] Tsyboulski, D. A.; Rocha, J. D. R.; Bachilo, S. M.; Cognet, L.; Weisman, R. B. *Nano Lett.* **2007**, *7*, 3080.

[104] Reich, S.; Thomsen, C.; Robertson, J. *Phys. Rev. Lett.* **2005**, *95*, 077402.

[105] Oyama, Y.; Saito, R.; Sato, K.; Jiang, J.; Samsonidze, G. G.; Gruneis, A.; Miyauchi, Y.; Maruyama, S.; Jorio, A.; Dresselhaus, G.; Dresselhaus, M. S. *Carbon* **2006**, *44*, 873.

[106] Journet, C.; Maser, W. K.; Bernier, P.; Loiseau, A.; delaChapelle, M. L.; Lefrant, S.; Deniard, P.; Lee, R.; Fischer, J. E. *Nature* **1997**, *388*, 756.

[107] Guo, T.; Nikolaev, P.; Thess, A.; Colbert, D. T.; Smalley, R. E. *Chem. Phys. Lett.* **1995**, *243*, 49.

[108] Thess, A.; Lee, R.; Nikolaev, P.; Dai, H. J.; Petit, P.; Robert, J.; Xu, C. H.; Lee, Y. H.; Kim, S. G.; Rinzler, A. G.; Colbert, D. T.; Scuseria, G. E.; Tomanek, D.; Fischer, J. E.; Smalley, R. E. *Science* **1996**, *273*, 483.

[109] Lebedkin, S.; Schweiss, P.; Renker, B.; Malik, S.; Hennrich, F.; Neumaier, M.; Stoermer, C.; Kappes, M. M. *Carbon* **2002**, *40*, 417.

[110] Jester, S. S.; Kiowski, O.; Lebedkin, S.; Hennrich, F.; Fischer, R.; Sturzl, N.; Hawecker, J.; Kappes, M. M. *Phys. Status Solidi B* **2007**, *244*, 3973.

[111] Jester, S. S. Nanostrukturierte Kohlenstoffverbindungen. Ph.D. Thesis, Institut für Physikalische Chemie, Universität Karlsruhe (TH), 2008.

[112] Bronikowski, M. J.; Willis, P. A.; Colbert, D. T.; Smith, K. A.; Smalley, R. E. *Journal of Vacuum Science & Technology a-Vacuum Surfaces and Films* **2001**, *19*, 1800.

[113] Nikolaev, P.; Bronikowski, M. J.; Bradley, R. K.; Rohmund, F.; Colbert, D. T.; Smith, K. A.; Smalley, R. E. *Chem. Phys. Lett.* **1999**, *313*, 91.

[114] Puretzky, A. A.; Geohegan, D. B.; Jesse, S.; Ivanov, I. N.; Eres, G. *Appl. Phys. A-Mater.* **2005**, *81*, 223.

[115] Baker, R. T. K. *Carbon* **1989**, *27*, 315.

[116] Wagner, R. S.; Ellis, W. C. *Appl. Phys. Lett.* **1964**, *4*, 89.

[117] Zhang, Y.; Li, Y.; Kim, W.; Wang, D.; Dai, H. *Appl. Phys. A-Mater.* **2002**, *74*, 325.

[118] Han, S. J.; Yu, T. K.; Park, J.; Koo, B.; Joo, J.; Hyeon, T.; Hong, S.; Im, J. *J. Phys. Chem. B* **2004**, *108*, 8091.

[119] He, M. S.; Duan, X. J.; Wang, X.; Zhang, J.; Liu, Z. F.; Robinson, C. *J. Phys. Chem. B* **2004**, *108*, 12665.

[120] Li, W. Z.; Xie, S. S.; Qian, L. X.; Chang, B. H.; Zou, B. S.; Zhou, W. Y.; Zhao, R. A.; Wang, G. *Science* **1996**, *274*, 1701.

[121] Ren, Z. F.; Huang, Z. P.; Xu, J. W.; Wang, J. H.; Bush, P.; Siegal, M. P.; Provencio, P. N. *Science* **1998**, *282*, 1105.

[122] Fan, S. S.; Chapline, M. G.; Franklin, N. R.; Tombler, T. W.; Cassell, A. M.; Dai, H. J. *Science* **1999**, *283*, 512.

[123] Hata, K.; Futaba, D. N.; Mizuno, K.; Namai, T.; Yumura, M.; Iijima, S. *Science* **2004**, *306*, 1362.

[124] Futaba, D. N.; Hata, K.; Yamada, T.; Mizuno, K.; Yumura, M.; Iijima, S. *Phys. Rev. Lett.* **2005**, *95*, 056104.

[125] Murakami, Y.; Chiashi, S.; Miyauchi, Y.; Hu, M. H.; Ogura, M.; Okubo, T.; Maruyama, S. *Chem. Phys. Lett.* **2004**, *385*, 298.

[126] Puretzky, A. A.; Eres, G.; Rouleau, C. M.; Ivanov, I. N.; Geohegan, D. B. *Nanotechnology* **2008**, *19*, 055605.

[127] Kong, J.; Soh, H. T.; Cassell, A. M.; Quate, C. F.; Dai, H. J. *Nature* **1998**, *395*, 878.

[128] Zhang, Y. G.; Chang, A. L.; Cao, J.; Wang, Q.; Kim, W.; Li, Y. M.; Morris, N.; Yenilmez, E.; Kong, J.; Dai, H. J. *Appl. Phys. Lett.* **2001**, *79*, 3155.

[129] Joselevich, E.; Lieber, C. M. *Nano Lett.* **2002**, *2*, 1137.

[130] Ismach, A.; Kantorovich, D.; Joselevich, E. *J. Am. Chem. Soc.* **2005**, *127*, 11554.

[131] Huang S., L. *J. Am. Chem. Soc.* **2003**, *125*, 5636.

[132] Huang, S. M.; Fu, Q.; An, L.; Liu, J. *Phys. Chem. Chem. Phys.* **2004**, *6*, 1077.

[133] Huang, S. M.; Maynor, B.; Cai, X. Y.; Liu, J. *Advanced Materials* **2003**, *15*, 1651.

[134] Huang, S. M.; Woodson, M.; Smalley, R.; Liu, J. *Nano Lett.* **2004**, *4*, 1025.

[135] Hong, B. H.; Lee, J. Y.; Beetz, T.; Zhu, Y. M.; Kim, P.; Kim, K. S. *J. Am. Chem. Soc.* **2005**, *127*, 15336.

[136] Jin, Z.; Chu, H. B.; Wang, J. Y.; Hong, J. X.; Tan, W. C.; Li, Y. *Nano Lett.* **2007**, *7*, 2073.

[137] Kiowski, O. Fluoreszenzmikroskopie an einzelnen Kohlenstoffnanoröhren. Diploma Thesis, Institut für Physikalische Chemie, Universität Karlsruhe (TH), 2004.

[138] Ibach, W.; Hollricher, O. Tutorial: High Resolution Optical Microscopy. WITec GmbH, 2002.

[139] Wong, S.; Kiowski, O.; Kappes, M.; Lindner, J.; Mandal, N.; Peiris, F. C.; Ozin, G. A.; Thiel, M.; Braun, M.; Wegener, M.; von Freymann, G. *Adv. Mat.,* submitted **2008**,

[140] Su, M.; Zheng, B.; Liu, J. *Chem. Phys. Lett.* **2000**, *322*, 321.

[141] de los Arcos, T.; Wu, Z. M.; Oelhafen, P. *Chem. Phys. Lett.* **2003**, *380*, 419.

[142] de los Arcos, T.; Garnier, M. G.; Oelhafen, P.; Mathys, D.; Seo, J. W.; Domingo, C.; Garci-Ramos, J. V.; Sanchez-Cortes, S. *Carbon* **2004**, *42*, 187.

[143] Harutyunyan, A. R.; Tokune, T.; Mora, E. *Appl. Phys. Lett.* **2005**, *87*, 051919.

[144] Kiowski, O.; Lebedkin, S.; Hennrich, F.; Malik, S.; Rösner, H.; Arnold, K.; Sürgers, C.; Kappes, M. M. *Phys. Rev. B* **2007**, *75*, 075421.

[145] Hennrich, F.; Krupke, R.; Arnold, K.; Stutz, J. A. R.; Lebedkin, S.; Koch, T.; Schimmel, T.; Kappes, M. M. *J. Phys. Chem. B* **2007**, *111*, 1932.

[146] Brintlinger, T.; Chen, Y. F.; Durkop, T.; Cobas, E.; Fuhrer, M. S.; Barry, J. D.; Melngailis, J. *Appl. Phys. Lett.* **2002**, *81*, 2454.

[147] Homma, Y.; Suzuki, S.; Kobayashi, Y.; Nagase, M.; Takagi, D. *Appl. Phys. Lett.* **2004**, *84*, 1750.

[148] Zhang, R. Y.; Wei, Y.; Nagahara, L. A.; Amlani, I.; Tsui, R. K. *Nanotechnology* **2006**, *17*, 272.

[149] Keldysh, L. V. *Phys. Status Solidi A* **1997**, *164*, 3.

[150] Perebeinos, V.; Tersoff, J.; Avouris, P. *Phys. Rev. Lett.* **2004**, *92*, 257402.

[151] Lefebvre, J.; Fraser, J. M.; Homma, Y.; Finnie, P. *Appl. Phys. A-Mater.* **2004**, *78*, 1107.

[152] Yin, Y.; Cronin, S.; Walsh, A.; Stolyarov, A.; Tinkham, M.; Vamivakas, A.; Bacsa, W.; Ünlü, M. S.; Goldberg, B. B.; Swan, A. K. *cond-mat/0505004* **2005**,

[153] Ohno, Y.; Iwasaki, S.; Murakami, Y.; Kishimoto, S.; Maruyama, S.; Mizutani, T. *Phys. Rev. B* **2006**, *73*, 235427.

[154] Okazaki, T.; Saito, T.; Matsuura, K.; Ohshima, S.; Yumura, M.; Iijima, S. *Nano Lett.* **2005**, *5*, 2618.

[155] Cronin, S. B.; Yin, Y.; Walsh, A.; Capaz, R. B.; Stolyarov, A.; Tangney, P.; Cohen, M. L.; Louie, S. G.; Swan, A. K.; Unlu, M. S.; Goldberg, B. B.; Tinkham, M. *Phys. Rev. Lett.* **2006**, *96*, 127403.

[156] Chakrapani, N.; Wei, B. Q.; Carrillo, A.; Ajayan, P. M.; Kane, R. S. *Proceedings of the National Academy of Sciences of the United States of America* **2004**, *101*, 4009.

[157] Kaatze, U. *Journal of Chemical and Engineering Data* **1989**, *34*, 371.

[158] Ellison, W. J.; Lamkaouchi, K.; Moreau, J. M. *Journal of Molecular Liquids* **1996**, *68*, 171.

[159] Sheng, C. X.; Vardeny, Z. V.; Dalton, A. B.; Baughman, R. H. *Phys. Rev. B* **2005**, *71*, 125427.

[160] Cognet, L.; Tsyboulski, D. A.; Rocha, J. D. R.; Doyle, C. D.; Tour, J. M.; Weisman, R. B. *Science* **2007**, *316*, 1465.

[161] Ausman, K. D.; Piner, R.; Lourie, O.; Ruoff, R. S.; Korobov, M. *J. Phys. Chem. B* **2000**, *104*, 8911.

[162] Sakai, H.; Suzuura, H.; Ando, T. *J. Phys. Soc. Jpn.* **2003**, *72*, 1698.

[163] Dukovic, G.; Wang, F.; Song, D. H.; Sfeir, M. Y.; Heinz, T. F.; Brus, L. E. *Nano Lett.* **2005**, *5*, 2314.

[164] Wang, Z. J.; Pedrosa, H.; Krauss, T.; Rothberg, L. *Phys. Rev. Lett.* **2006**, *96*, 047403.

[165] Tamor, M. A.; Wolfe, J. P. *Phys. Rev. Lett.* **1980**, *44*, 1703.

[166] Yin, Y.; Walsh, A. G.; Vamivakas, A. N.; Cronin, S.; Stolyarov, A. M.; Tinkham, M.; Bacsa, W.; Ünlü, M. S.; Goldberg, B. B.; Swan, A. K. *cond-mat/0606047* **2006**,

[167] Arnold, K.; Lebedkin, S.; Kiowski, O.; Hennrich, F.; Kappes, M. M. *Nano Lett.* **2004**, *4*, 2349.

[168] Basché, T.; Moerner, W. E.; Orrit, M.; Wild, U. P. *Single-Molecule Optical Detection, Imaging and Spectroscopy*; VCH, Weinheim, 1997.

[169] Moerner, W. E.; Orrit, M. *Science* **1999**, *283*, 1670.

[170] VandenBout, D. A.; Yip, W. T.; Hu, D. H.; Fu, D. K.; Swager, T. M.; Barbara, P. F. *Science* **1997**, *277*, 1074.

[171] Empedocles, S.; Bawendi, M. *Accounts of Chemical Research* **1999**, *32*, 389.

[172] Mason, M. D.; Credo, G. M.; Weston, K. D.; Buratto, S. K. *Phys. Rev. Lett.* **1998**, *80*, 5405.

[173] Geddes, C. D.; Parfenov, A.; Gryczynski, I.; Lakowicz, J. R. *Chem. Phys. Lett.* **2003**, *380*, 269.

[174] Efros, A. L.; Rosen, M. *Phys. Rev. Lett.* **1997**, *78*, 1110.

[175] Frantsuzov, P. A.; Marcus, R. A. *Phys. Rev. B* **2005**, *72*, 155321.

[176] Htoon, H.; O'Connell, M. J.; Cox, P. J.; Doorn, S. K.; Klimov, V. I. *Phys. Rev. Lett.* **2004**, *93*, 027401.

[177] Hennrich, F.; Krupke, R.; Lebedkin, S.; Arnold, K.; Fischer, R.; Resasco, D. E.; Kappes, M. *J. Phys. Chem. B* **2005**, *109*, 10567.

[178] Hartschuh, A.; Pedrosa, H. N.; Novotny, L.; Krauss, T. D. *Science* **2003**, *301*, 1354.

[179] Matsuda, K.; Kanemitsu, Y.; Irie, K.; Saiki, T.; Someya, T.; Miyauchi, Y.; Maruyama, S. *Appl. Phys. Lett.* **2005**, *86*, 123116.

[180] Lefebvre, J.; Finnie, P.; Homma, Y. *Phys. Rev. B* **2004**, *70*, 045419.

[181] Naumov, A. V.; Bachilo, S. M.; Tsyboulski, D. A.; Weisman, R. B. *Nano Lett.* **2008**, *8*, 1527.

[182] Wang, F.; Dukovic, G.; Knoesel, E.; Brus, L. E.; Heinz, T. F. *Phys. Rev. B* **2004**, *70*, 241403.

[183] Vijayaraghavan, A.; Kar, S.; Soldano, C.; Talapatra, S.; Nalamasu, O.; Ajayan, P. M. *Appl. Phys. Lett.* **2006**, *89*, 162108.

[184] Seferyan, H. Y.; Nasr, M. B.; Senekerimyan, V.; Zadoyan, R.; Collins, P.; Apkarian, V. A. *Nano Lett.* **2006**, *6*, 1757.

[185] Zhu, Z. P.; Crochet, J.; Arnold, M. S.; Hersam, M. C.; Ulbricht, H.; Resasco, D.; Hertel, T. *Journal of Physical Chemistry C* **2007**, *111*, 3831.

[186] Mortimer, I. B.; Nicholas, R. J. *Phys. Rev. Lett.* **2007**, *98*, 027404.

[187] Berger, S.; Voisin, C.; Cassabois, G.; Delalande, C.; Roussignol, P.; Marie, X. *Nano Lett.* **2007**, *7*, 398.

[188] Kane, C. L.; Mele, E. J. *Phys. Rev. Lett.* **2004**, *93*, 197402.

[189] Lefebvre, J.; Homma, Y.; Finnie, P. *Phys. Rev. Lett.* **2003**, *90*, 217401.

[190] Jeong, G. H.; Yamazaki, A.; Suzuki, S.; Yoshimura, H.; Kobayashi, Y.; Homma, Y. *Carbon* **2007**, *45*, 978.

[191] Kiowski, O.; Lebedkin, S.; Kappes, M. M. *Phys. Status Solidi B* **2006**, *243*, 3122.

[192] Abe, M.; Kataura, H.; Kira, H.; Kodama, T.; Suzuki, S.; Achiba, Y.; Kato, K.; Takata, M.; Fujiwara, A.; Matsuda, K.; Maniwa, Y. *Phys. Rev. B* **2003**, *68*, 041405.

[193] Freitag, M.; Tsang, J. C.; Kirtley, J.; Carlsen, A.; Chen, J.; Troeman, A.; Hilgenkamp, H.; Avouris, P. *Nano Lett.* **2006**, *6*, 1425.

[194] Duesberg, G. S.; Loa, I.; Burghard, M.; Syassen, K.; Roth, S. *Phys. Rev. Lett.* **2000**, *85*, 5436.

[195] Duan, X. J.; Son, H. B.; Gao, B.; Zhang, J.; Wu, T. J.; Samsonidze, G. G.; Dresselhaus, M. S.; Liu, Z. F.; Kong, J. *Nano Lett.* **2007**, *7*, 2116.

[196] Breeden, K. H.; Langley, J. B. *Review of Scientific Instruments* **1969**, *40*, 1162.

[197] Inoue, T.; Matsuda, K.; Murakami, Y.; Maruyama, S.; Kanemitsu, Y. *Phys. Rev. B* **2006**, *73*, 233401.

[198] Hertel, T.; Perebeinos, V.; Crochet, J.; Arnold, K.; Kappes, M.; Avouris, P. *Nano Lett.* **2008**, *8*, 87.

[199] Gao, B.; Duan, X. J.; Zhang, J.; Wu, T. J.; Son, H. B.; Kong, J.; Liu, Z. F. *Nano Lett.* **2007**, *7*, 750.

[200] Yang, L.; Anantram, M. P.; Han, J.; Lu, J. P. *Phys. Rev. B* **1999**, *60*, 13874.

[201] Yang, L.; Han, J. *Phys. Rev. Lett.* **2000**, *85*, 154.

[202] Cherukuri, P.; Bachilo, S. M.; Litovsky, S. H.; Weisman, R. B. *J. Am. Chem. Soc.* **2004**, *126*, 15638.

[203] Leeuw, T. K.; Reith, R. M.; Simonette, R. A.; Harden, M. E.; Cherukuri, P.; Tsyboulski, D. A.; Beckingham, K. M.; Weisman, R. B. *Nano Lett.* **2007**, *7*, 2650.

[204] Cherukuri, P.; Gannon, C. J.; Leeuw, T. K.; Schmidt, H. K.; Smalley, R. E.; Curley, S. A.; Weisman, R. B. *Proceedings of the National Academy of Sciences of the United States of America* **2006**, *103*, 18882.

[205] Tsyboulski, D. A.; Bachilo, S. M.; Weisman, R. B. *Nano Lett.* **2005**, *5*, 975.

List of Publications

[1] K. Arnold, S. Lebedkin, <u>O. Kiowski</u>, F. Hennrich, M. M. Kappes: "Matrix-imposed stress-induced shifts in the photoluminescence of single-walled carbon nanotubes at low temperatures", *Nano Letters* **2004**, *4*, 2349-2354.

[2] J. T. Fermann, T. Moniz, <u>O. Kiowski</u>, T. J. McIntire, S. M. Auerbach, T. Vreven, M. J. Frisch: "Modeling proton transfer in zeolites: Convergence behavior of embedded and constrained cluster calculations", *Journal of Chemical Theory and Computation* **2005**, *1*, 1232-1239.

[3] <u>O. Kiowski</u>, K. Arnold, S. Lebedkin, F. Hennrich, M. M. Kappes: "Detection of single carbon nanotubes in aqueous dispersion via photoluminescence", *AIP conference proceedings* **2005**, *786*, 139-142.

[4] <u>O. Kiowski</u>, S. Lebedkin, M. M. Kappes: "Photoluminescence microscopy of as-grown individual single-walled carbon nanotubes on Si/SiO2 substrates", *Physica Status Solidi B-Basic Solid State Physics* **2006**, *243*, 3122-3125.

[5] S. Lebedkin, K. Arnold, <u>O. Kiowski</u>, F. Hennrich, M. M. Kappes: "Raman study of individually dispersed single-walled carbon nanotubes under pressure" *Physical Review B* **2006**, *73*, 094109.

[6] S. S. Jester, <u>O. Kiowski</u>, S. Lebedkin, F. Hennrich, R. Fischer, N. Stürzl, J. Hawecker, M. M. Kappes: "Combination of atomic force microscopy and photoluminescence microscopy for the investigation of individual carbon nanotubes on sapphire surfaces", *Physica Status Solidi B-Basic Solid State Physics* **2007**, *244*, 3973-3977.

[7] O. Kiowski, K. Arnold, S. Lebedkin, F. Hennrich, M. M. Kappes: "Direct observation of deep excitonic states in the photoluminescence spectra of single-walled carbon nanotubes", *Physical Review Letters* **2007**, *99*, 237402.

[8] O. Kiowski, S. Lebedkin, F. Hennrich, M. M. Kappes: "Single-walled carbon nanotubes show stable emission and simple photoluminescence spectra with weak excitation sidebands at cryogenic temperatures", *Physical Review B* **2007**, *76*, 075422.

[9] O. Kiowski, S. Lebedkin, F. Hennrich, S. Malik, H. Rösner, K. Arnold, C. Sürgers, M. M. Kappes: "Photoluminescence microscopy of carbon nanotubes grown by chemical vapor deposition: Influence of external dielectric screening on optical transition energies", *Physical Review B* **2007**, *75*, 075421.

[10] S. Lebedkin, F. Hennrich, O. Kiowski, M. M. Kappes: "Photophysics of carbon nanotubes in organic polymer-toluene dispersions: Emission and excitation satellites and relaxation pathways", *Physical Review B* **2008**, *77*, 165429.

[11] S. Wong, O. Kiowski, M. Kappes, J. Linder, N. Mandal, F. C. Peiris, G. A. Ozin, M. Thiel, M. Braun, M. Wegener, G. von Freymann: "Spatially localized photoluminescence at 1.5 micrometers wavelength in direct laser written optical nanostructures", *Advanced Materials* **2008**, *20*, 4097-4102.

[12] O. Kiowski, S. S. Jester, S. Lebedkin, Z. Jin, Y. Li, M. M. Kappes: "Photoluminescence Spectral Imaging of Ultra-Long Single-Walled Carbon Nanotubes on Silicon: Micromanipulation Induced Strain, Torsion, Rupture and the Determination of Handedness", *submitted to Nano Letters*.

Acknowledgements

An dieser Stelle möchte ich mich bei allen Personen bedanken, die mich unterstützt, und zum Gelingen dieser Arbeit beigetragen haben.

- Prof. M. M. Kappes für die Möglichkeit, in seinem Arbeitskreis zu promovieren und die mir zugestandenen Freiheiten.

- Dr. Sergei Lebedkin für die Betreuung meiner Arbeit, die Hilfe bei praktischen und theoretischen Problemen und für die Geduld und nützlichen Tipps.

- Dr. Sharali Malik für das Korrekturlesen dieser Arbeit und die TEM Aufnahmen von CVD Nanoröhren.

- Frank Hennrich für die Bereitstellung von Nanoröhrendispersionen und die Messung der Durchmesserverteilung der CVD Röhren.

- Stefan Jester für die Zusammenarbeit mit der Nanoröhrenstatistik und dem Verbiegen von SWNTs.

- Prof. Yan Li und Zhong Jin für die Kooperation und die Bereitstellung von ihren CVD Nanoröhren.

- Katharina Arnold, Marco Neumaier, Mattias Kordel, Anne Lechtken, Ninette Stürzl und Lars Walter für die gute Atmosphäre, Diskussionen und Hilfe.

- Allen weiteren Mitgliedern des Arbeitskreises und Mitarbeitern des INT, vor allem Thomas Koch und Ralph Krupke für die Bereitstellung von AFM, SEM und der Sputter-, Spincoating- und Ionenätzanlage.

- Allen Mitarbeitern und Kollegen des Arbeitskreises am Lehrstuhl II der Universität Karlsruhe.

Die VDM Verlagsservicegesellschaft sucht für wissenschaftliche Verlage abgeschlossene und herausragende

Dissertationen, Habilitationen, Diplomarbeiten, Master Theses, Magisterarbeiten usw.

für die kostenlose Publikation als Fachbuch.

Sie verfügen über eine Arbeit, die hohen inhaltlichen und formalen Ansprüchen genügt, und haben Interesse an einer honorarvergüteten Publikation?

Dann senden Sie bitte erste Informationen über sich und Ihre Arbeit per Email an *info@vdm-vsg.de*.

Sie erhalten kurzfristig unser Feedback!

VDM Verlagsservicegesellschaft mbH
Dudweiler Landstr. 99
D - 66123 Saarbrücken
www.vdm-vsg.de

Telefon +49 681 3720 174
Fax +49 681 3720 1749

Die VDM Verlagsservicegesellschaft mbH vertritt

Printed by Books on Demand GmbH, Norderstedt / Germany